Animal Welfare

Series Editor

Professor Clive Phillips
Foundation Chair of Animal Welfare
Centre for Animal Welfare and Ethics
School of Veterinary Science
University of Queensland
Gatton 4343, QLD
Australia

For further volumes:
http://www.springer.com/series/5675

Terry L. Maple • Bonnie M. Perdue

Zoo Animal Welfare

Springer

Terry L. Maple
Wilkes Honors College
Biology and Psychology Departments
Florida Atlantic University
Jupiter/Boca Raton, FL
USA

Bonnie M. Perdue
Department of Psychology
Agnes Scott College
Atlanta, GA
USA

ISSN 1572-7408
ISBN 978-3-642-35954-5 ISBN 978-3-642-35955-2 (eBook)
DOI 10.1007/978-3-642-35955-2
Springer Heidelberg New York Dordrecht London

Library of Congress Control Number: 2013933027

Printed on acid-free paper

Springer is part of Springer Science+Business Media (www.springer.com)

Foreword

Throughout the Western world a debate is taking place on the morality of confinement. On one side are the preservationists and reformers who see existing conditions of human confinement in institutions such as mental hospitals, orphanages, and prisons as satisfactory (preservationists) or in need of improvement (reformers) and on the other side are the abolitionists who view confinement itself as immoral and unethical. These divergent views inform current discussion regarding the purposes and the morality of the modern zoo. Improving zoo conditions and enhancing their educational value, two important goals of reformers, will not mute the objections of those who see animal confinement itself as unethical and immoral. This is why discussion of zoo animal welfare is so important. The book uses welfare in its original *positive* sense, from the Middle English "a well-faring" or wish for someone to have a good journey, avoiding the current negative image that has motivated so many welfare departments to change their names to social services departments.

The book begins with the senior author's insider account of the turnaround and revitalization of Zoo Atlanta through evidence-based reform. The next chapter lists the challenges in defining and measuring animal welfare. Because conditions found in nature are not necessarily suitable for captive animals, the book recommends the use of multiple measures with continuing assessment of outcomes. Animal preference can be one criterion but there are times when it is an unreliable guide to animal welfare.

Chapter 3 discusses research design, especially the use of operant conditioning and physiological measures in zoo research. Data collection and analysis must be suited to zoo realities, which often means small samples of selected populations monitored by volunteer student assistants. Chapter 4 introduces the concept of wellness, which takes the discussion beyond well-being into the domain of positive psychology. It was nice to see this orientation complementing Skinner's behavioristic psychology. Chapter 5 describes the value of academic training in animal behavior and the need for research partnerships between zoos and universities which will bring faculty and students into zoos and field stations.

Chapter 6 (Environmental Enrichment) describes methods for adding sensory stimulation and providing choice and challenge. Because these enrichment efforts are not always successful, there will be a need for species-specific assessment.

Chapter 7 discusses behavioral analysis and training using the technology derived from Skinner's operant conditioning and schedules of reinforcement. The authors suggest recruiting behavior therapists using this approach successfully with humans to advise zookeepers, zoo biologists, and zoo veterinarians.

Chapter 8 summarizes the latest developments in zoo design and shows how field observations can be combined with critical evaluation of zoo practice. The authors are familiar with recent developments in zoo design both nationally and internationally. Optimal zoo size is covered as is the debate between advocates of comprehensive collections and those emphasizing particular regions or species. The section on "Encouraging constructive criticism," develops a major theme of the book, the value of bringing together zoo designers and practitioners with their critics in workshops and symposia. Two people whose views are discussed in detail are David Hancocks, former zoo director and now strident critic of many zoo practices, and Jon Coe, the innovative and futurist zoo designer.

Chapter 9 (Launching Ethical Arks) pulls it all together. Maple and Perdue believe that providing optimal animal welfare will "pay off" in terms of visitor numbers, satisfaction, and support. Active, fit, and healthy animals attract the public and give them greater satisfaction. The authors describe Dr. Maple's leadership experience in developing partnerships between zoos, animal welfare organizations, and universities. This chapter considers several relevant antinomies (conflicts between valid approaches) using the example of debates over conservation practices with endangered species. Both sides have valid positions and only research and discussion can produce workable solutions. The empirical zoo can be the staging ground for innovation and change. The book ends discussing the Comparative Quality of Life, a holistic approach that accounts for the unique perspectives, preferences, and needs of individuals.

Good sailing Ethical Ark. May debate and fruitful outcomes set your course.

Robert Sommer
Professor Emeritus
University of California at Davis

Animal Welfare Series Preface

Animal welfare is attracting increasing interest worldwide, especially in developed countries where the knowledge and resources are available to, at least potentially provide better management systems for farm animals, as well as companion, zoo and laboratory animals. The key requirements for adequate food, water, a suitable environment, companionship and health are important for animals kept for all of these purposes.

There has been increased attention given to animal welfare in the West in recent years. This derives largely from the fact that the relentless pursuit of financial reward and efficiency, to satisfy market demands, has led to the development of intensive animal management systems that challenge the conscience of many consumers in this part of the world, particularly in the farm and laboratory animal sectors. Livestock are the world's biggest land users (FAO 2002) and the farmed animal population is increasing rapidly to meet the needs of an expanding human population. This results in a tendency to allocate fewer resources to each animal and to value individual animals less, for example in the case of farmed poultry where flocks of over 20,000 birds are not uncommon. In these circumstances, the importance of each individual's welfare is diminished.

In developing countries, human survival is still a daily uncertainty, so that provision for animal welfare has to be balanced against human welfare. Animal welfare is usually a priority only if it supports the output of the animal, be it food, work, clothing, sport or companionship. However, in many situations the welfare of animals is synonymous with the welfare of the humans that look after them, because happy, healthy animals will be able to assist humans best in their struggle for survival. In principle the welfare needs of both humans and animals can be provided for, in both developing and developed countries, if resources are properly husbanded. In reality, the inequitable division of the world's riches creates physical and psychological poverty for humans and animals alike in many parts of the world.

Increased attention to welfare issues is just as evident for zoo, companion, laboratory, sport and wild animals. Of growing importance is the ethical management of breeding programs, since genetic manipulation is now technically advanced, but there is less public tolerance of the breeding of extreme animals if it comes at the expense of animal welfare. The quest for producing novel genotypes has fascinated breeders for centuries. Dog and cat breeders have produced a variety of deformities that have adverse effects on their welfare, but nowadays the breeders are just as active in the laboratory, where the mouse is genetically manipulated with equally profound effects.

The intimate connection between animals and humans that was once so essential for good animal welfare is rare nowadays, having been superseded by technologically efficient production systems where animals on farms and in laboratories are tended by increasingly few humans in the drive to enhance labour efficiency. With today's busy lifestyles, companion animals too may suffer from reduced contact with humans, although their value in providing companionship, particularly for certain groups such as the elderly, is beginning to be recognized. Consumers also rarely have any contact with the animals that are kept for their benefit.

In this estranged, efficient world, people struggle to find the moral imperatives to determine the level of welfare that they should afford to animals within their charge. A few people, and in particular many companion animal owners, strive for what they believe to be the highest levels of welfare provision, while others, deliberately or through ignorance, keep animals in impoverished conditions in which their health and well-being can be extremely poor. Today's multiple moral codes for animal care and use are derived from a broad range of cultural influences, including media reports of animal abuse, guidelines on ethical consumption and campaigning and lobbying groups.

This series has been designed to contribute towards a culture of respect for animals and their welfare by producing learned treatises about the provision for the welfare of the animal species that are managed and cared for by humans. The early species-focused books were not detailed management blue-prints; rather they described and considered the major welfare concerns, often with reference to the behavior of the wild progenitors of the managed animals. Welfare was specifically focused on animals' needs, concentrating on nutrition, behavior, reproduction and the physical and social environment. Economic effects of animal welfare provision were also considered where relevant, as were key areas where further research is required.

In this volume the series again departs from the species focus to address animals in zoos. Few areas of animal management have attracted more controversy over the last 50 years, with zoo animals' welfare, conservation value, ability to entertain and role in educating the public being evaluated in a prolonged debate as to whether it is ethical to keep animals in zoos. People's position in this debate depends usually on the relative value that they place on these possible roles that zoo animals can play. Professor Terry Maple has had a lifetime's experience with zoos and is a major campaigner and educator for improved animal welfare in zoos. As Director of Zoo Atlanta and a former wildlife psychologist, Dr. Maple has reshaped many American zoos into models for zoos around the world, using his belief in naturalistic design and a strong sense of purpose for the modern zoo. That sense of purpose comes across strongly in this inspirational volume, Zoo Animal Welfare.

In it Maple and Perdue raise the intriguing possibility of zoos having a major role as conservators of fauna and flora of the local area in which they are located. For the sceptics I'd say "read it, and then tell me you are still sceptical about zoos", for the believers in zoos having a purpose in the modern world, I'd say, "be prepared to be inspired by this book".

Food and Agriculture Organization (2002). http://www.fao.org/ag/aga/index_en.htm.

St. Lucia, QLD
Australia

Clive Phillips

Preface

In graduate school at the University of California at Davis, I had the good fortune to work in a research lab in developmental psychobiology investigating the consequences of social deprivation. I learned quickly that disruptions in primate affectional systems, especially the mother-infant bond, inevitably induced psychopathology. As my interest in zoos developed, I was surprised to discover an entire generation of monkeys and apes exhibiting the familiar signs and symptoms of social isolation. Normally socialized primates and other social mammals are nurtured by attentive mothers (and sometimes fathers) providing the stimulation necessary for species-appropriate social and cognitive development. A functional infancy is an absolute requirement for optimal animal welfare. Studying a group of the original isolate monkeys (*Macaca mulatta*) produced in the Wisconsin primate lab of Harry F. Harlow, I made a decision to devote my career to the restoration of natural social networks and the prevention of psychopathology in zoo animals.

Captivity and Behavior (1979), co-edited with Joe Erwin and Gary Mitchell, was an early opportunity to survey the domain of an emerging field of psychological research, at the same time producing our first comprehensive contribution to the literature of nonhuman primate welfare. It contained one of the first contributions to primate enrichment, a benchmark chapter written by Hal Markowitz. For four decades, my research group has benefited from the thorough and exhaustive research of Professor Harlow and his collaborators, many of whom graduated to found productive primate labs of their own. One of Harlow's students, Gary Mitchell, was my research mentor at Davis. He encouraged me to explore the zoo as a research setting, and later sent other young scientists to probe this goldmine of scientific opportunity (Mitchell et al. 1991). Decades of experimental research on social deprivation in rhesus monkeys provided the necessary insight into developmental disorders in a variety of nonhuman primate species living in zoos. Vilified by animal rights groups, Harlow and his academic family is actually responsible for the documentation and discovery that inspired better standards and management practices for primates in captive settings (e.g. Blum 2002). Two of his scientific protégées generated the benchmark publication that elevated psychological well-being to the forefront of animal welfare science (Novak and Suomi 1988).

An equally powerful driver of animal welfare is the physical environment. Robert Sommer's classic paper "What do we learn at the zoo?" demonstrated how inadequate zoo environments influenced the attitudes and beliefs of visitors. Deprivation acts and stereotyped behaviors in animals coping with conditions of social deprivation are exacerbated by barren physical facilities. Sommer suggested the label "hard architecture" to classify environments bereft of naturalistic features. By contrast, "soft architecture" encourages social interaction, exploration, and activity. Soft space is by its nature flexible, comfortable, and user-friendly. These principles apply to animals and people alike. Beyond the zoo, Sommer identified hospitals, prisons, and airports as examples of hard architecture. Unfortunately, as Professor Sommer revealed in his iconic book *Tight Spaces* (1974), hard architecture has become the uncomfortable norm for all of us.

Early in my academic career I worked with my students to formulate a behavioral model for creating functional animal habitats in the zoo (Maple and Stine 1982; Maple and Finlay 1986, 1987). My forays into zoo exhibit psychology paralleled the aesthetic design revolution that resulted in the visionary landscapes at Woodland Park Zoo in Seattle. A developing partnership with the zoo designers Gary Lee and Jon Coe facilitated the intellectual framework that resulted from a continuing conversation now 35 years and counting. Zoo Atlanta and CLR ascended to prominence in tandem, the result of a harmonious and highly creative partnership in the psychology of design. Our crucible of new ideas and innovations were thoroughly vetted in the classroom during three decades of teaching an interdisciplinary course in "Psychology and Environmental Design" in the College of Architecture at Georgia Tech. I am grateful to Dr. Jean Wineman, now at the University of Michigan, who was my teaching partner and research collaborator for many productive years.

My ongoing solidarity with zoo keepers, curators, and veterinarians who have suffered the frustration of hard labor in substandard facilities, encouraged me to write books to encourage innovation and document the change. *Orang-utan Behavior* (Maple 1980a) and *Gorilla Behavior* (Maple and Hoff 1982) anticipated the concept of an "empirical zoo" (Maple and Lindburg 2008). My research group at Emory University and Georgia Tech responded to the need for information and new ideas by publishing dozens of papers in peer-reviewed journals on a wide variety of species. Along the way, we had the opportunity to organize a new journal to facilitate the emerging science of zoo biology. As we worked together to shape the new discipline, I am grateful for decades of helpful feedback I've received in an active correspondence with zoo colleagues. Their questions and suggestions have helped us to formulate better research projects, solve real problems, and reach an audience that has become truly international in scope. These friendly relationships gave me access behind-the-scenes in dozens of zoos and aquariums throughout the world. A career long friendship and scholarly partnership with Joe Erwin provided the motivation to generate social change through entrepreneurial leadership.

Zoo Animal Welfare has been written to encourage significant change in zoos, aquariums and similar institutions, and to engender a culture of respect for animals as envisioned by the editors who established the Animal Welfare series at Springer.

As the reader will quickly discover the book is more zoo than aquarium oriented. Welfare is an issue common to both and our ideas are applicable to both. Of course, zoos today often contain major aquatic exhibits, and aquariums are increasingly comfortable exhibiting birds, reptiles, marine mammals, and even large felids and ursids. The breadth of biodiversity in zoos and aquariums demands an astute and comprehensive understanding of welfare.

As a scholar in academia and a decision maker in zoos, I have enjoyed the unique perspective of one who could actually put my ideas to the test. In Atlanta, I led an organization that experienced for two decades nothing short of revolutionary change, and the institutional commitment to animal welfare is still working for my successors and former collaborators. In my opinion, zoo animal welfare works on many levels. It is a very strong marketing concept validating the organization's commitment to maintaining a healthy population of zoo animals. "Animal care," a common euphemism for welfare, is growing in importance as zoos strenuously compete for the support of their communities. Presently and long into the future, zoos that are known for their commitments to conservation and animal welfare will surely grow and prosper.

The zoos and aquariums of the future will be designed with welfare in mind, and they will provide the tools and the context to approach if not achieve a state of optimal animal welfare. Indeed, a shift to the priority of welfare has already begun. The fact that organized, accredited zoos in Europe are fully committed to zoo animal welfare is encouraging to those of us who work in the United States. We are not there yet, although there is broad agreement that we are moving in the right direction. The first step is to acknowledge the elevated priority of animal welfare, and to make the adjustments in programs and personnel to enact the change. Directors must lead the change, and ultimately work with their communities to fund the change. Welfare-oriented exhibits can be costly but their impact on the animals will be appreciated by every zoo visitor and easily justify the expenditure. The priority of conservation is needed to save wildlife; the elevated priority of welfare will ensure the survival of zoos and aquariums dedicated to protecting wildlife.

As the leader of institutions that sought to extend the reach of welfare and the science that sustains it, I understand how challenging it is to introduce and monitor substantive, even radical changes in facilities, programs, and operational routines. Optimal animal welfare requires big ideas that enable zoo animals to live large, long and well. Dr. Perdue and I have dedicated *Zoo Animal Welfare* to the universe of zoo professionals who work each day to provide the best possible life for the animals entrusted to their care, and to the current generation of students in colleges and universities who harbor a passion to someday work with exotic fauna in the zoo. In addition, we have dedicated this book to celebrating the life and legacy of our good friend and mentor, Hal Markowitz, who died while this book was in preparation.

Recent collaborations at the San Francisco Zoo have strengthened my appreciation for the zoo keeper's essential role in advancing the health and welfare of zoo animals. Keepers, curators, and veterinarians have been especially important in helping us to formulate our ideas about wildlife wellness and welfare. In San Francisco, David Bocian, Graham Crawford, and Joe Fitting have been extremely helpful in the development of the wildlife wellness concept that we introduced in this book. I thank Chairman, David Stanton; CEO, Tanya Peterson and the Board of Directors of the San Francisco Zoo for encouraging and supporting our work.

Springer Series Editor Dr. Clive Phillips generously provided a detailed review of the manuscript, and we are grateful to other colleagues and friends for their helpful comments on selected chapters. Anette Lindqvist patiently and carefully managed the project from start to finish. Her encouragement was restorative at strategic moments in the long process of writing a book.

Bonnie and I are especially grateful for our collaboration and academic kinship with a continuous cadre of brilliant graduate and undergraduate students who have worked with us in California, Georgia, and Florida and at distant field sites in Africa and China. Without their efforts and profound insights this book would not be possible.

Zoos and Aquariums have made great strides in our lifetime, but as good as they have become, we believe they are still operating well below their full potential. Our ultimate success requires candor and critical thinking. Without an ongoing, objective self-appraisal, we cannot become credible advocates for the animals in our care. We trust that *Zoo Animal Welfare* will be regarded as a bold step in this direction.

Jupiter, FL Terry L. Maple, Ph.D.

About the Authors

Terry L. Maple is Research Professor of Biology and Psychology, and Affiliate Professor at the Harriett Wilkes Honors College of Florida Atlantic University. He is also Professor-in-Residence at the San Francisco Zoo and a leadership consultant to non-profit organizations. He previously served as the President/CEO of Zoo Atlanta and the Palm Beach Zoo. Dr. Maple was elected president of the Association of Zoos and Aquariums in 1999. For a decade he served as the Elizabeth Smithgall Watts Professor in the School of Psychology at the Georgia Institute of Technology where he founded and directed the Center for Conservation & Behavior. Dr. Maple served for 4 years on the board of the U.S. Institute for Museum and Library Services, a Presidential appointment. As the Founding Editor he launched the scientific journal *Zoo Biology* in 1982. Dr. Maple and his many students and collaborators have written and edited more than 200 scientific books and papers, including *Ethics on the Ark* (1995) and *Great Apes and Humans: the Ethics of Coexistence* (2001) both published by Smithsonian Institution Press.

Bonnie M. Perdue is a member of the faculty at Agnes Scott College. Her postdoctoral research was conducted at the Language Research Center at Georgia State University in Atlanta. She received her Ph.D. in Cognition and Brain Sciences with an emphasis on Animal Behavior and Comparative Psychology from the School of Psychology at the Georgia Institute of Technology where she served as the Coordinator of the Center for Conservation & Behavior. Dr. Perdue has conducted behavioral research on giant pandas in China, otters, elephants, flamingos, and a variety of non-human primate taxa at Zoo Atlanta and the Language Research Center at Georgia State University. Her papers have been published in many peer-reviewed journals including *Biology Letters*, the *Journal of Comparative Psychology*, the *Journal of Applied Animal Welfare Science*, and *Zoo Biology*.

Contents

Chapter 1
Building Ethical Arks

> *Zoos have the marvelous potential to develop a concerned, aware, energized, enthusiastic, caring, and sympathetic citizenry. Zoos can encourage gentleness toward all other animals and compassion for the well-being of wild places . . . To help save all wildlife, to work toward a healthier planet, to encourage a more sensitive populace; these are the goals for the new zoos.*
>
> David Hancocks

The aspirations enumerated by our colleague David Hancocks reveal the awesome potential of the world's best zoos and aquariums. Similar words and phrases can be found in the mission statements of a growing number of accredited institutions. One eloquent and visionary statement of purpose broke new ground when it was issued in 1980 by leaders at the Minnesota Zoo: "Strengthening the bond between people and the living earth". Recently, the zoo modified its mission statement to read: "Connecting people, animals, and the natural world". In San Francisco, the zoo lives by the motto: "Connect, care, conserve". Such elegant phraseology frames each and every institutional commitment to ethical principles, core values, and superior operating standards and practices. On the ethical ark, the words matter.

In a comprehensive review of ethics in the zoo profession, Kreger and Hutchins (2010) took the position that ethics is about "what is right and what is wrong. Further, they argued, rather than focusing on *what is*, as scientists do, ethicists are concerned with what *ought to be* (White 1981). The mere fact that we have chosen to capture exotic fauna and deposit them in our zoos is an example of an "ethical paradox" as Conway (1995) explained it:

> Zoos seek to inspire public interest in wild creatures and nature, to provide ecological education, and to help save wild species from extinction, but in doing so they confine wild animals away from nature and manage their lives. (p. 2)

Because so many animals live in world zoos (more than 750,000 are estimated to reside in the world's accredited zoos), it is essential that we consider their welfare and

T.L. Maple and B.M. Perdue, *Zoo Animal Welfare*, Animal Welfare,
DOI 10.1007/978-3-642-35955-2_1, © Springer-Verlag Berlin Heidelberg 2013

understand our ethical obligation to keep them healthy and well. Both government and private enterprise have recognized the growing importance of ethical operating principles, ethical decision-making, and ethical commitments. Specialized courses in ethics are now among the most important electives in our universities and graduate schools of business, law, and public policy. Non-profit organizations are particularly concerned about ethics given their reliance on funding from local and national foundations, corporations, and individuals.

1.1 Fall and Rise of the Phoenix

The collapse of Atlanta's city zoo in 1984 was a scandal that embarrassed government officials and reverberated throughout the nation. In an article in *Parade Magazine*, the Humane Society of the United States named the Atlanta Zoo as one of America's ten worst zoos. Due to a series of well-publicized (Desiderio 2000) ethical lapses including the secretive translocation of a dying elephant discovered buried in a shallow ditch in Cherryville, North Carolina, the Association of Zoos and Aquariums (AZA) discontinued Atlanta's membership. Twinkles the elephant became a national symbol of irresponsible zoo management and firmly positioned the Atlanta Zoo as the profession's number one pariah (Fig. 1.1). Fortunately, Atlanta's business and government leaders acted decisively to reverse the zoo's misfortune, implementing and funding in 1985 a bold plan for its revitalization. Rebranded as "Zoo Atlanta" the zoo was restructured as a non-profit corporation, owned by the City of Atlanta, but operated independently by a nine-member board of private business leaders. Time and again the new zoo board made decisions demonstrating that its first priority was the health and welfare of the zoo population. Its total institutional transformation brought the zoo into alignment with the highest standards of America's most accomplished zoological parks. Zoo Atlanta received AZA accreditation in 1987, just 3 years after the peak of its crisis.

By 1989, *Parade Magazine* was singing Atlanta's praises as the city that had turned its zoo around. Atlanta's experience proved that any zoo can overcome a substandard operating history if it commits to advancing the health and welfare of the animal collection. The rebirth of Atlanta's zoo, a virtual Phoenix story, was AZA's first successful privatization and a triumph of ethics over institutional inertia. Former Mayor Andrew Young proclaimed that Zoo Atlanta was the most successful public-private partnership in Georgia history, and it has been recognized nationally as a model conversion of public to private governance. Beginning his zoo life isolated and confined to a dilapidated steel, concrete, and tile cage, the lowland gorilla Willie B. ended his life in a simulated, landscaped jungle, a venerable silverback with his own harem of females and a collection of offspring (Maple 2001). He had become the symbol and brand for a revitalized Zoo Atlanta. New zoo exhibits were naturalistic in form and function, designed to encourage natural behavior, breeding, and normal parenting in gorillas and other species. As Wineman and Choi (1991) discovered, the 1985 zoo master-plan represented a complete shift in design philosophy when

Fig. 1.1 Doing hard time at the Atlanta Zoo c. 1984 (R.D. Fowlkes)

compared to the 1950s era plan. The earlier plan prioritized entertainment, whereas the reform plan put the welfare of animals first. While the 1950s plan was clearly detrimental to animals, the new plan worked for animals and people, although visitors were now situated at a greater distance from them.

In his moving eulogy for Willie B. who died on February 3, 2000, Mayor (and Reverend) Young observed (Fig. 1.2):

> We looked at him (Willie B.) in his cage, and we knew that he didn't belong there. He was brought here in captivity but he found a way to appeal to our hearts so that we were moved to find ways to set him free. And in setting him free, perhaps we set ourselves free to help us learn that we can live together in peace with all of the animals that God has created.

Since 1985, there have been many organizational setbacks experienced by zoos in America and around the world, including the global malaise that followed 9–11, and the catastrophic global recession in 2008. As debilitating as a financial crisis can be, no challenge compares to the crisis in public confidence that follows a disastrous animal incident. The death of an employee from attack by an elephant, whale, or leopard, an attack on a visitor by an escaped gorilla, tiger, or bear, or the loss of a beloved animal under mysterious circumstances is uncommon, but serious mistakes or errors of judgment can shut down any institution, or force dramatic changes in the way it can be operated. We believe that attention to the ethical foundation of zoos and aquariums serves to inoculate them against mismanagement and the public hysteria that often accompanies catastrophic events. Operating reforms are preferable *before* a crisis occurs, and should be implemented by proactive audits from boards, management, and outside experts who can provide systematic, objective evaluations of the facility, its operating units, and the health, safety, and welfare of the zoo population. In Atlanta, a Technical Advisory Board was established soon after the non-profit corporation was formed, comprised of area veterinarians, scientists, and the President of the Atlanta Humane Society, with the responsibility to provide advice on animal welfare and ethical practices to the CEO. Many zoos such as the Smithsonian National Zoo have

Fig. 1.2 Willie B.'s first day outdoors after 27 years in a cage (J. Sebo)

established permanent animal welfare committees to monitor their progress and identify urgent problems. Some are developing sensitive evaluation instruments to facilitate routine audits of exhibits, practices, and standards. An emergency can become a mandate for changing a substandard institution, but it is frequently the case that mediocrity, once accepted, is the immediate precursor to catastrophic failure.

1.2 Debating Our Critics

Zoo Atlanta's commitment to ethical operating standards provided new opportunities for national leadership. In 1992 the zoo's scientific partnership with the Georgia Institute of Technology led the two institutions to host a workshop funded by the National Science Foundation where zoo and aquarium leaders entered into a constructive dialogue with a cadre of environmental ethicists and scholarly advocates of animal welfare and animal rights. The conference was sponsored by the Association of Zoos and Aquariums, a decision that our co-editor Ben Beck regarded as a "risky, pioneering effort to establish a scholarly dialogue." The aim of the conference was to examine current zoo practices and standards, debate unsettled and controversial issues, and discover (if possible) common ground for future cooperation, collaboration, and reform. Since its publication by Smithsonian Institution Press in 1995, *Ethics on the Ark* has been widely disseminated to an international audience of reform-minded zoo professionals. The book's subtitle, "Zoos, Animal Welfare, and Wildlife Conservation" is no accident, as conservation and animal welfare were regarded by the participants as compatible and synergistic. One of the most important collective recommendations from the Atlanta conference was the need for animal welfare groups, zoos, and aquariums to work together to improve humane standards in substandard facilities. Perhaps the most hotly debated issue at the workshop was the question of whether zoos and aquariums willingly subordinated individual animal welfare to the higher priority of populations (species). As a champion of the rights of individuals, the philosopher Dale Jamieson (1995) provided an analogy in the arrogance of totalitarian governments:

> Vaclav Havel has said that the fall of communism has lessons not just for people in the East but for people everywhere. Communism, according to Havel, was the ultimate human attempt to manage everything. It failed because it could manage nothing. Our attempts to manage nature are, if anything, more arrogant than communist attempts to manage human societies. We need some humility, and to recognize that our attempts to make things better may well make things worse. (p. 72)

In North America, the zeal of AZA Species Survival Plan (SSP) committees during their formative years certainly made it clear to curators and directors that breeding decisions must be determined by group data and the metrics of population biology. The firmly established commitment to the new strategy of animal management didn't leave much room to debate the psychological consequences of moving animals around the country, and more than one recommendation was resisted by local community adversaries. A few high-profile transactions such as the recommendation that the Cleveland Zoo gorilla Timmy be translocated for breeding purposes to the Bronx Zoo in New York ended in court (Timmy turned out to be a very successful breeder in New York.). In many cases, animal rights advocates responded to the contention that pair-bonded primates would unduly suffer from the act of separation. One newspaper issued the headline: "Zoo to dissolve orangutans' 22-year pair bond," offering a longtime donor's opinion that the move was "extreme cruelty" even though orangutans do not live in pairs in the wild. It turned out, however, that zoo biologists serving on management committees were themselves increasingly concerned about demographically dictated translocations that might compromise individual animal welfare. What the Atlanta meeting clearly demonstrated was the efficacy of *both* population and animal welfare priorities. In response to welfare concerns, SSP committees are obliged to take into account the social and behavioral consequences of moving animals for breeding purposes, and acknowledge that some delicate species and certain individuals are particularly difficult to move. Social history and social bonds might be sufficient reasons to minimize translocations in some species. Today, no rational critic would suggest that zoo professionals are dismissive of individual animal welfare. Indeed, A commitment to managed conservation programs such as the AZA Species Survival Plan does not preclude superior animal welfare, but the commitment to both priorities is not easy or inexpensive.

1.3 Ethical Obligations to Welfare

The World Association of Zoos and Aquariums (WAZA) published a revision of its benchmark *World Zoo and Aquarium Conservation Strategy* in 2005. The document included a new section on "Ethics and Animal Welfare". In this section the authors asserted that "zoos and aquariums have a moral obligation to contribute to the conservation of habitats and biodiversity in the interests of society and of the animals themselves." Further, they recognized that the future of zoos and aquariums would depend on ethical justifications for maintaining living collections, and a strong commitment to the welfare of the animals in their care. WAZA leadership acknowledged the fluctuating sentiments of visitors and media while accepting the value of critical thinking, deliberation, and debate. The looming parity of conservation and welfare is suggested by the following conclusion in WAZA's strategy:

> All zoos and aquariums must be widely trusted as caretakers of animals and focus not only on the ultimate goal of conservation, but also on meeting the immediate needs of the living creatures for which they are responsible. (pp. 59–60)

Although ethics permeates all dimensions of zoo management, zoo animal welfare is fundamentally an ethical issue. An ethical ark, therefore, may be defined as "a zoo or aquarium that is committed to advancing superior animal welfare standards and practices." An institution that is fully committed to animal welfare will search exhaustively for welfare deficiencies using the scientific assessment instruments we have described in this book. The search will only be successful if there are sufficient human resources trained and empowered to discover institutional shortcomings.

1.4 Who Monitors Welfare in the Zoo?

Ethical arks are built, in part, by assigning institutional and personal responsibility for systematically monitoring the welfare of individual animals. Certainly keepers and curators, on the front lines of animal management, and their veterinary and vet-tech colleagues offer dedicated professional care, husbandry, and treatment to each and every individual. However, even if zoo animal welfare has risen to a position of higher priority among other priorities, it has not achieved this level of acceptance in all zoos and aquariums. Greater accountability may require changes in organizational structure, for example designating responsibility for animal welfare to key personnel whose only or primary responsibility is monitoring the behavior of zoo animals. Elsewhere (Maple 2007) we have advocated the recruitment of resident animal behavior specialists who are accountable for objective welfare policies, standards, and practices at the zoo. Historically, zoos have been prone to de-emphasize assignments of this kind, settling for employees with lesser credentials. For example, a survey by Anderson et al. (2010) found that only 30 % of the staff responsible for research had earned doctoral degrees. It has been our experience that personnel with Ph.D.'s in animal behavior are capable of functioning at a much higher level of impact and authority, and therefore worth the small difference in compensation. Possessing the proper academic credentials they are also well-equipped to network with other doctoral level professionals in the scientific community and in nearby universities, and they are qualified to serve as mentors to other staff. Just one doctoral level professional on a zoo or aquarium staff can elevate the management conversation on a daily basis.

Accredited zoos and aquariums need to operate well ahead of their critics, anticipating problems and influencing exhibit design and daily operating routines. Indeed, they should be adept at constructive self-criticism, empirical monitoring, and proactive intervention. The zoo architect Gary Lee, a founding partner in the Philadelphia-based design firm CLR, has likened animal management to software, arguing that superior hardware (exhibits) cannot be successful without superior software, that is, sound and attentive management practices. Even the most exceptional exhibits have to be intelligently managed to reach their full operating potential. In addition, although zoos and aquariums have embraced new management techniques such as environmental enrichment, zoo managers rarely have the time or the training to evaluate them or to publish their findings in reputable professional journals. In contrast, doctoral training in anthropology, ethology or comparative psychology enables the designated animal behavior specialist to systematically observe the animal's response to new architecture or new interventions. The behavioral scientist's statistical acumen provides for evaluations that are objective and meaningful (and publishable).

A dedicated behavioral scientist with an operant orientation can interact with trainers to get the most out of schedules of reinforcement and automated feeding systems. Key scientific personnel can systematically monitor behavior, expand animal behavior literacy, and mentor others to achieve a more holistic understanding of the creatures under our care. Broadly applied, proactive scientific management identifies problems and attempts to resolve them before they develop into controversies. A scientist on staff is not just obligated to conduct research. He or she must also be empowered to help peers, e.g. curators, veterinarians, and keepers, to gather, organize, and evaluate data for dissemination. For example, the quantitative skills of a senior graduate student assigned to work with an experienced reptile curator at Zoo Atlanta contributed to the successful publication of years of data from field observations of American alligators (Hunt and Ogden 1991). Doctoral level managers at scientifically grounded North American institutions such as Brookfield, Bronx, Lincoln Park, and San Diego have contributed time-saving instruments for evaluation that retain keepers in the role of observers, as recently described by Canino and Powell (2010). With a keeper-friendly data system in place, the "multi-point scans" method introduced by Margulis et al. (2005), zookeepers were able to evaluate their polar bear enrichment program without disrupting their normal routine.

We believe that personnel trained in animal behavior are just as important to a zoo or aquarium as medical, educational, or curatorial specialists. Indeed, as experts in the measurement and interpretation of behavior, they provide a valuable source of information germane to all aspects of animal health and welfare. Since most zoos and aquariums do not employ staff with expertise in behavior, it is no wonder that our institutions have difficulty fighting back against the credentialed adversaries frequently recruited to the debate by our most strident critics. The science of animal behavior crosses all boundaries and connects all disciplines in the zoo, and may be the key to significantly raising the bar to achieve the highest possible quality of zoo animal welfare. Behavioral scientists are also qualified to join curators in their intellectual defense against "anthropomorphism by omission" or the failure to consider that other animals experience the world differently. As the comparative psychologist Gordon Burghardt explained it:

> We can, without realizing it, attribute human traits to other species by failing to consider that many animals perceive the world in a different manner . . . different species have different perspectives and priorities . . . Rivas and Burghardt (2002) (p. 10)

For more than 30 years, zoos and aquariums worldwide have become increasingly scientific, but they have not yet reached a critical mass of intellectual capital (Maple and Bashaw 2010). Small and medium sized institutions must recognize that it is not costly to build a foundation of expertise in scientific zoo biology through partnerships and it is clearly in their best interest to do so. Institutions can be rapidly transformed by investments of this kind, enabling more proactive and profound contributions to an expanding culture of scientific animal welfare. Simply by setting up a lecture series, smaller zoos can bring new ideas into their realm, and often the cost can be shared by a nearby university. In consecutive years, the annual "Conservation leadership Lecture Series" at the Palm Beach Zoo brought field biologists Alan Rabinowitz, George Schaller, and field veterinarian Billy Karesh to speak to large audiences of local supporters and spend quality time with zoo staff, board members, and their guests. Zoo animal welfare and welfare science will be firmly established once enough emissaries have spread the word. Another way to enrich the intellectual climate of a zoo or aquarium is to seed the institutions with talented graduate students. When there is a consistent pipeline of talent through collaboration and mentoring, an empirical transformation is inevitable, and management has an important human resource to support evidence-based decisions.

A key to advancing zoo animal welfare is the development and testing of creative innovations. Often the idea can be found in the iconic publications and public statements of scientific pioneers such as Robert M. Yerkes or Heini Hediger. Long before there were objective standards for animal welfare, Yerkes and Hediger knew that captivity could be dramatically improved, and they sought to provide animals with far better living standards. Each wrote thoughtful treatises on the wildlife they studied in nature, the zoo, and the laboratory. Sometimes innovations derive from surprising sources such as the humanistic psychology of Abraham Maslow. In considering his ideas on self-actualization and human potential, we discovered psychological principles that, with a little tweaking, applied to nonhuman primates and other animals. In Chap. 5 we will discuss a perspective on animal welfare based on Maslow's ideas. His characterization suggested animals were innately predisposed to live a good and natural life.

Another approach to achieve greater accountability for institutional animal welfare is the classification of "animal welfare" as a working job title. To our knowledge, the Detroit Zoo is the only North American zoo committed to recruiting dedicated animal welfare staff. A search of the open positions advertised in the September issue of the AZA *Connect* magazine, reveals dozens of employment opportunities, but not one job with "welfare" in its title. However, in 2010, the Detroit Zoo authorized two staff positions, Director of Animal Welfare, and Animal Welfare Manager, to drive the agenda of their innovative Center for Zoo Animal Welfare. Two years later, as this book goes to press, Detroit is in the process of filling the senior position after the departure of their first Director of Animal Welfare. The original ad that was circulated by the Detroit Zoo in 2009 demonstrates the comprehensive nature of their program:

> The Director of Animal Welfare develops, conducts, facilitates and oversees applied research aimed at assessing and improving the welfare of zoo animals and contributing to knowledge about captive exotic animal welfare; consults and cooperates with curators and veterinarians in animal husbandry and health care management; oversees programs for environmental enrichment and operant conditioning training; and contributes to the development of zoo animal welfare philosophy and policy.

Their unique Center for Zoo Animal Welfare was created by Detroit Zoo leadership to accomplish the following institutional objectives:

1. To operate as a resource center to organize and disseminate knowledge, research, and best animal welfare practices;
2. To function as a forum and a convener for exotic animal welfare science, practice and policy discussions;
3. To conduct research and training in zoo animal welfare and recognize advances in the field.

In 2011 a national workshop for animal caretakers and welfare leaders was organized by the Detroit Zoological Society with the compelling title, "From Good Care to Great Welfare". The meeting brought together a diverse group of professionals from universities, zoos, aquariums, humane societies, and animal welfare organizations from North America, Europe, and the United Kingdom. Wayne Pacelle, President of the Humane Society of the United States (HSUS) delivered the keynote address. In many ways the conference complemented and expanded on the topics first explored at the Atlanta ethics workshop. The key feature of both meetings was the commitment to finding common ground among individuals and organizations with a history of disagreement and discord. These two gatherings of experts and practitioners clearly demonstrated that there is plenty of opportunity for cooperation and collaboration to advance animal welfare science, applications, practices, standards, and policy. Mr. Pacelle's address spoke eloquently of the need for accredited zoos and HSUS to cooperate in setting higher standards for animals living in captive settings. He acknowledged that AZA differentiated itself from so-called "roadside attractions" in part by the adherence of AZA member institutions to a code of professional ethics and a demanding accreditation program. There can be no doubt that a unified front to upgrade or eliminate roadsides would be a great achievement by HSUS in collaboration with AZA, but there is no sign that this idea is gaining traction.

It appears that a dynamic AZA-HSUS partnership will require more time, better communication, and greater trust. A follow-up workshop hosted by the Detroit Zoological Society in 2012 was organized to share a framework that can be implemented to advance a zoo animal welfare agenda. In addition, the forum was designed to suggest design principles that will guide the assembly of essential components, practices, actions, techniques, and measures that contribute to animal well-being in zoos and aquariums (Fig. 1.3). The Detroit Center for Animal Welfare is asserting its national leadership while providing a visible venue to debate and discuss animal welfare issues with stakeholders in zoos, aquariums, universities, and allied humane organizations.

Zoos and aquariums seeking to elevate the priority of welfare are not obligated to authorize new welfare positions. They can simply add welfare to the responsibilities of current employees, but welfare should not be an afterthought. In many institutions in Europe and in North America, welfare practices and standards have been delegated to veterinary or curatorial staff. In some cases, research personnel have assumed responsibility for animal welfare. Scientists in the zoo can recruit student research assistants to monitor individual animal welfare. Although keepers can be trained to gather data, their jobs are so structured that they often have little time to observe and record. Systematic data provide feedback to keepers and animal managers, but also provide information that can be shared with peers through publication. Research guides the implementation of welfare standards and practices.

In San Francisco, zoo leadership recently reclassified a position in senior management to recognize the importance of "wellness" as an institutional priority. As we hope to demonstrate in Chap. 4, the concept of wellness should be regarded as synonymous with welfare. The examples reviewed in Chap. 4 demonstrates how zoos and aquariums are assigning responsibility and accountability for animal welfare throughout the organization. Optimal animal welfare/wellness has become a major strategy driving the global zoo vision.

Three decades of government regulation has driven operating standards and practices in Europe, but recent criticism from groups like the Born Free Foundation have raised questions about the efficacy of these standards. In response, officials of the British and Irish Zoological Association (BIAZA) reported that 83 % of the 192 zoos rated by government met the recommended standards. Further, they noted that member institutions performed significantly better than non-members of BIAZA (Daily Mail online, October 9, 2012). Similarly, the World Zoo and Aquarium Association acknowledged the need for accountability and for objective evaluation of animal welfare in all zoological institutions in their membership. There is no ambiguity in their message:

> The ethical and welfare issues involved in managing wild animals in collections need to be constantly addressed and evaluated. This is essential for the future of zoos and aquariums and for their ability to implement their core missions of conservation, education and science. Such goals will only be met when zoos and aquariums have the trust of their visitors and donors. (p. 63)

Fig. 1.3 Detroit Zoo exhibits demonstrate an institutional commitment to animal welfare

1.5 Good to Great Welfare

Jim Collins' (2001) best-selling book, *Good to Great*, resonated throughout the business community, while his companion primer, *Good to Great for the Social Sectors*, is a highly relevant and useful guide for operating zoos and aquariums. Collins acknowledges the fundamental truth that social sector organizations must focus on long-term performance relative to mission and not just financial returns. It's not that financial issues are unimportant, but when an organization delivers on its mission it is likely to be successful on all fronts including the business side. A singular focus on business principles may bring prosperity but not greatness, according to Collins. If we apply this principle to welfare, a zoo or aquarium committed to great welfare will satisfy the needs of its living collection of wildlife and the people who pay to see them. Good welfare is inherent in the operating philosophy of all successful zoos, but great welfare requires an extraordinary commitment throughout the organization. We believe that zoo visitors expect and want the organization to designate the health and welfare of the animals as its first priority.

European experts have generated useful metrics for evaluating welfare, and their productivity and impact is measured by publications in a variety of journals including those that have developed to advance animal welfare in agriculture and biomedicine. A curator concerned with the welfare of exotic birds will discover a plethora of agricultural journals that are broadly concerned with avian husbandry techniques. Great welfare requires a commitment to its broad scientific foundation. The empirical zoo concept is one manifestation of this commitment (Maple and Lindburg 2008). One could argue that empirical zoos are more likely to generate knowledge that is critical to advancing from good to great. With the growing number of scientist-practitioners in the curatorial domain, welfare is now recognized as a primary responsibility of zoo leaders and managers.

The five-year strategic plan of the Association of Zoos and Aquariums is one of the documents that elevated animal welfare to become a key priority for accredited zoos in North America. AZA charges the association and its members to: "Foster the advancement and implementation of animal care and welfare methods and standards for use by AZA members as best practices in maintaining healthy animal collections." In 2005, AZA established an Animal Welfare Committee to, among other priorities, assist members to continually improve the welfare of animals in their care. The committee of qualified scientists, curators, and veterinarians from AZA institutions and their academic allies is impressive. One of the committee's first steps produced a working definition of animal welfare (Barber and Mellen 2008):

> Animal welfare is the degree to which an animal can cope with challenges in its environment as determined by a combination of measures of health (including pre-clinical physiological responses) and measures of psychological well-being. (p. 41)

Zoo professionals throughout the world recognize there is much progress to be made through innovations in animal care including the profession's lengthy history of creative environmental enrichment. Other associations have expressed their commitments on websites around the world. The British-Irish Association of Zoos and Aquariums (BIAZA) statement on animal welfare is straightforward:

> The promotion of good welfare for zoo animals is a priority which BIAZA takes very seriously. Zoos and Aquariums meet the needs of the animals in their care by understanding what constitutes good welfare, and by providing appropriate housing and husbandry.

Research is clearly an important component of zoo animal welfare, but some of our critics do not fully comprehend the responsibilities of an empirical zoo. According to Jamieson (1985):

> The fact that zoo research contributes to improving conditions in zoos is not a reason for having them. If there were no zoos, there would be no reason to improve them. (p. 113)

But zoos not only exist, they are ubiquitous, and as Hutchins et al. argued in rebuttal, zoo studies have also provided important information to help protect animals in the wild. For both reasons, we believe, research in zoos must be encouraged. Jamieson (1985) has also written that zoos can be justified on the basis of their educational mission, so he is at best an inconsistent critic. When there is sufficient critical mass in our scientific units, we are capable of great leaps forward. Recently, AZA institutions coalesced to carry out a comprehensive institutional study of elephant welfare. The organizers recruited the participation of 49 North American zoos currently exhibiting elephants. Supported by funding from the US Institute for Museum and Library Services (IMLS), this ongoing project is gathering important data about elephant management in AZA zoos. Suitably, the project chair is a well-published scholar and zoo director, Anne Baker. Large research collaborations such as this one are becoming commonplace, the result of unprecedented cooperation at the level of regional and national associations. Much of the heavy lifting is assigned to scientific curators. With many AZA institutions reserving the right to utilize aversive control methods, it will be interesting to see if the report directly confronts the negative consequences of these practices.

We have argued that zoos and aquariums need dedicated scientific staff positions to advance and monitor welfare, but one problem with the current scientist-practitioner model is the inevitable erosion of research time as the responsibilities of management exert their priority. At all levels, managers with both scientific and curatorial responsibilities have reported difficulty in fulfilling their scientific duties. Scientific curators who cannot express their full competence are likely to experience frustration and they will not reach their full potential as zoological leaders. Such an outcome renders the zoo less impacting on a national or international level. Executive leadership is required to keep these organizational responsibilities in balance. Otherwise, as Hediger warned us, science is relegated to a lower priority, and with it goes welfare.

1.6 Institutional Leadership

The Detroit Zoo's effort to elevate animal welfare builds on scientific momentum disseminated through scholarly journals and major conferences. The benchmark 1993 conference on environmental enrichment hosted by the Metro Washington Park Zoo in Portland, Oregon (now the Oregon Zoo) generated enthusiasm and energized keepers, curators and veterinarians. The main papers delivered at this meeting appeared in *Second Nature*, an important book (Shepherdson et al. 1998) published by the Smithsonian Press. This was the second volume in a series that started with *Ethics on the Ark*. The Portland tradition of applied behavioral research has been ably passed from Markowitz to Mellon to Shepherdson, a very good example of the power of scholars to envision, plan, and implement innovation and reform in zoos and aquariums. Thanks to pioneering institutions such as the Oregon Zoo, enrichment has become an institutional core value in every accredited zoo and aquarium throughout the world.

Chicago's Brookfield Zoo asserted their global leadership in 2008 by establishing a center for animal welfare research and hosting an international symposium on the subject. Originally conceived as a center dedicated to the science of "animal well-being", Brookfield leaders wisely decided to brand the unit to encompass greater breadth and it was subsequently named the "Center for the Science of Animal Welfare" (Dan Wharton 2012, personal communication). The symposium attracted an impressive network of scholars and curators. Selected papers from this conference appeared in a special issue of *Zoo Biology* (2009). The Brookfield center is focused on identifying best practices of animal care and builds on the zoo's strong science-based programs including "population genetics, behavioral endocrinology, nutrition, behavioral husbandry, behavioral research, and veterinary science." (Website, Chicago Zoological Society). With so much good work on welfare science emanating from the Brookfield and Lincoln Park Zoos, Chicago could be accurately identified as the North American locus for zoo animal welfare science. Each of the institutions that have taken the lead in animal welfare offers a slightly different approach, so there is ample opportunity for both collaboration and specialization. The esteemed John G. Shedd Aquarium, also located in Chicago, represents an opportunity to spread the welfare concept to cognitively complex aquatic animals, such as seals and sea lions, dolphins, and Beluga whales. Their extensive use of operant conditioning training techniques represents an excellent bridge to innovations in aquatic animal welfare. Shedd Aquarium also hosts specialized training programs for other zoo and aquarium professionals. There is great potential at Shedd Aquarium to advance aquatic standards of animal welfare if they can strengthen the connection between the science and practice of behavior analysis and training. We should look to the world's most esteemed aquariums and ocean parks to stimulate innovation in aquatic animal management. We'll have more to say about this issue in Chap. 7.

1.7 Conservation and Welfare

Both AZA and WAZA regard welfare as a component of conservation. Clearly, a commitment to comprehensive conservation measures would demand a focus on welfare. In captivity the focus is on living well, but in nature survival is the first priority. We believe it is time to consider giving equal institutional priority to conservation and welfare, and to acknowledge that zoos and aquariums can pay homage to both. Innovative field partnerships are showing how both conservation and welfare can be served through the relentless protection of species and their ecosystems. The work of Panthera is a case in point.

Based in New York City, Panthera (www.panthera.org) is widely recognized as the leading great cat conservation organization in the world. It is a non-governmental organization working in partnership with the Wildlife Conservation Society and other NGO's on the ground in Asia, Africa, North America, and South and Central America. Together they have promoted innovative conservation action plans. One such plan, "Tigers Forever" offers unprecedented accountability to donors while pledging to increase Asian tiger populations in select locations by 50 % in 10 years. A highly ambitious field strategy, tigers are monitored by observers but also through innovative field technology. Throughout their range, tiger movements can be tracked through the use of camera-traps. When a beam of light is broken by a passing tiger, an image of the tiger (or any other animal) is recorded. By comparing the photos to computer records the individual animal can be identified and its position calculated. Camera traps have also been deployed to study the movements of jaguars in Colombia (Fig. 1.4).

Through an active monitoring process and by police or military intervention to prevent poachers from gaining the upper hand, great cat populations can be protected so they can successfully reproduce. Active protection can also be directed to the forest ecosystem that sustains tigers since poachers who kill smaller animals deprive tigers of the food sources they require. The goal is to protect thriving ecosystems occupied by a sustainable population of great cats and other diverse fauna. Survival is therefore the ultimate form of welfare, and both tigers and their prey benefit from intervention by park rangers and Panthera's local field assistants who continuously monitor at-risk

Fig. 1.4 Jaguar image captured with camera trap in Colombia (Courtesy Panthera)

populations in the field. Given the terribly brutal acts committed by poachers, wildlife morbidity and mortality are issues that conservationists must confront in the field. By protecting tigers and other wildlife from poachers, we protect them from the cruel use of wire snares and other vicious tools that produce a slow, painful death. Poachers use advanced technology to outcompete predators for bushmeat, killing with automatic weapons and operating from airborne helicopters. They kill tigers to obtain their skins, bones, and organs to satisfy the illicit trade in wildlife products. Beyond mere survival, wildlife welfare equivalent to a state of physical and psychological well-being is impossible while poachers are free to practice their brutal craft. Some of the world's most charismatic mega-vertebrates have been driven to the brink of extinction by organized poachers. In Africa, rhinos can be protected only when they are confined in reserves and private enclaves surrounded by electric fences and armed guards. Migrating elephants are more vulnerable, but both rhinos and elephants suffer the brutality of mutilation when criminals assault them to steal their valuable horns and tusks. Poachers are so bold that they recently broke into a high security zoo in Itanagar, India and killed and mutilated a captive tiger. The problem is so serious that the government of India permits wildlife authorities to shoot and kill armed poachers on sight.

The fact that African elephants are being killed at a rate approaching 40,000 animals per year means that the species could be extinct by the year 2025. In just 1 year, 2006, 11 metric tons of illegal ivory were seized from ships bound for Asian countries such as China, Japan, and Taiwan. Another example of the dramatic losses is Zakouma National Park in Chad where more than 3,000 elephants were lost in a 4 year period, reducing their population to just 600 animals. In that same period 11 forest rangers were also killed by determined poachers. For those who believe that zoos should not exhibit elephants and for other critics who oppose captive breeding programs, it is quite possible that zoos will be the last refuge for a species on the verge of extinction in the wild. This is another good reason to build superior facilities for herds of zoo elephants that might become a reservoir for the species, and ultimately an opportunity for re-introduction of captive bred stock. Given these facts, the zeal and the urgency of regional associations to breed elephants is a rational response to the looming crisis.

Another dimension of wildlife welfare is the application of veterinary medicine to combat disease in wild populations. Animals with tuberculosis and other dangerous diseases often suffer a slow and painful death, and animals in pain represent a distinct danger to nearby human populations. Field veterinarians support conservation projects by treating sick and injured animals in the field. To this end, snares embedded in the skin of mountain gorillas, if removed in time, can prevent infection or amputation. Tuberculosis is transmitted when wild predators consume tainted domestic cattle. In the Ngorongoro Crater in Tanzania a resident population of lions was decimated when they developed distemper, a virus normally confined to canines. In Africa, Asia, Central and South America, veterinarians have intervened to vaccinate domestic animals that may infect wild lions, tigers, jaguars and other predatory cats. If welfare is defined as a reduction in suffering then conservation medicine may be regarded as a form of wildlife welfare.

As we have seen, many of the handling and medical procedures pioneered for exotic animals in the zoo have enabled veterinarians to successfully capture, restrain, and treat sick and injured wild animals in the field. New medical technologies and new drugs are rendering these interventions less invasive with rapid recovery. A growing number of zoos are offering the services of their keepers, curators, and veterinarians to participate in high priority, collaborative conservation medicine partnerships throughout the world. Welfare ideas and methods are widely shared by zoo biologists who use them in the zoo and in the field. Modern studies of nutrition benefit from coordinated work in the wild where a comprehensive knowledge of diets contributes to our understanding of biodiversity (Jordan 2005). In this way, our emerging welfare science is directly contributing to wildlife conservation while fieldwork contributes to animal welfare in the zoo.

Many conservation organizations work closely with zoos and aquariums. The Dian Fossey Gorilla Fund International (DFGFI) is headquartered on the campus of Zoo Atlanta (www.gorillafund.org). The synergy between Zoo Atlanta's highly regarded scientific program and its focus on lowland gorillas provides a capable partner for DFGFI, an organization operating field projects in Central Africa for mountain gorillas and the *Grauri* subspecies.

DFGFI scientists and their African tracking units have operated continuously despite the danger of war and tribal genocide throughout the region. In 2010 DFGFI founded a new operational unit known as the Gorilla Rehabilitation and Conservation Center (GRACE). This unique center can accommodate up to fifteen gorillas living in natural social groups with room to roam over 350 acres of natural habitat adjacent to the Tayna Nature Reserve in Congo. Orphaned gorillas rescued from poachers are undergoing both medical and psychological rehabilitation at GRACE but the center is also an important educational facility for the region, helping the local people to understand the importance of wildlife, protected parks and reserves, and the need to intervene whenever the welfare of wild gorillas is compromised by poachers or other

Fig. 1.5 Orphaned Grauer's gorillas en route to GRACE (Dian Fossey Gorilla Fund International)

encroachments. Zoos and zoo field partners are directly involved in the support of GRACE with sustaining contributions of money and resources from the Walt Disney Company, DFGFI, and their conservation partner Zoo Atlanta (Fig. 1.5).

The availability of other partners in the region provides a critical mass of expertise and human resources. One organization, the Mountain Gorilla Veterinary Project, is comprised of a team of veterinarians who are working to improve the sustainability of wild mountain gorilla populations through health monitoring, lifesaving veterinary care, relevant health studies, and the dissemination of knowledge gained in the field. Clearly, the work of the MGVP is an example of advancing animal welfare through the application of veterinary medicine. Many of the animals treated by MGVP vets have been injured by snares from local poachers, or by encroaching military units engaged in the regional civil war. Welfare is synonymous with conservation in Africa, and humane intervention benefits both individuals and populations.

Many of the veterinarians who work in Africa for MGVP are zoo veterinarians from North America and Europe, who willingly donate their time to assist in protecting and treating endangered gorillas in the field. In this way, they are also helping to establish a conservation and welfare ethic to sustain wildlife populations whose survival depends ultimately on the good will of local African people. Of course the vigilent protection of charismatic wildlife also protects the region's access to revenue from ecotourism, the lifeblood of many small communities. The health of surrounding communities in one of Africa's most densely populated regions also receives attention from MGVP and its partner organization Gorilla Doctors. Gorilla Doctors is affiliated with the University of California at Davis Wildlife Health Center. Other zoo partners include the Maryland Zoo where Mike Cranfield the co-director of Gorilla Doctors is affiliated, and Zoo Boise which recently contributed a grant to support orphan gorilla care at the MGVP field site. Maryland Zoo, located in the city of Baltimore, contributes space and facilities to store tissue and fluid samples collected from gorillas in the field and offices and laboratories for affiliated veterinarians such as Dr. Cranfield. The scope of the partnerships sustaining this program is impressive.

Fig. 1.6 Treatment of a mountain gorilla's snare-damaged foot (Dian Fossey Gorilla Fund International)

The Gorilla Doctors and DFGFI work within the context of the "One Medicine" approach to field conservation. This idea recognizes that human and veterinary medicine are closely connected, especially on the conservation front lines of Africa. Veterinarians and scientists working for both organizations attempt to manage disease in gorillas and nearby human villages. For a decade DFGFI has worked in partnership with the Bisate Health Center near the Volcanoes National Park to conduct an annual de-worming campaign in nearby villages like Bisate. This program originated to protect mountain gorillas from human disease. Intestinal parasites are a significant and recurring threat to both humans and gorillas due to a lack of safe drinking water. Thus, Fossey Fund employees are directly involved in preventing disease by attention to clean water sources, adequate sanitation, and public education. A new partnership with Emory University's Rollins School of Public Health is providing these communities with additional services from world class experts in parasite control (Fig. 1.6).

The most pressing conservation problem in West and Central Africa is the tragic business of bushmeat hunting, but this trade has now expanded into traditional tourist destinations in East and South Africa (Lindsay et al. 2012). While many iconic species are hunted for their skins, tusks, and horns, other animals starve because their traditional prey species are severely depleted by poaching for human consumption. The great cats are often victims of snares meant for smaller animals. Snares are highly effective, difficult to control, and result in widespread suffering and death for predatory cats and other animals. The tremendous growth in human populations in Africa is blamed for the increased demand for bushmeat. For many animals, cheetahs and African wild dogs for example, their presence outside protected parks makes them particularly vulnerable. Snares are insidious threats to Africa's populations of lions. In Mozambique's *Niassa* Reserve over half of the lion mortalities are due to snares.

The World Association of Zoos and Aquariums (WAZA) could be an effective political instrument to mobilize governments to fight and defeat the bushmeat trade.

Fig. 1.7 Trader's Alley at Zoo Atlanta teaches visitors about the illegal trade in wildlife products (A. Thompson)

Many zoos are already educating their visitors even though the reality of bushmeat is a disturbing topic. Zoo Atlanta's graphic exhibit (Fig. 1.7) on the global illegal wildlife trade is a step in the right direction. In an evaluation of "Trader's Alley" Perdue and Stoinski (unpublished) found that the exhibit improved the short and long-term knowledge of zoo visitors. In addition, zoo visitors who studied the exhibit graphics and text agreed that illegal wildlife trade was a significant problem and understood that citizen action was necessary to combat the trade.

Stoinski et al. (2002) investigated the exposure of zoo visitors to explicit photos of animals killed in the bushmeat trade. Visitors exposed to the photos agreed with the statement that zoos should endeavor to teach about the bushmeat problem. It appears that a zoo, largely positioned as a happy family destination, can be an effective venue for delivering unvarnished, alarming conservation messages. The bushmeat trade also illustrates the close link between conservation and animal welfare. Ethical arks in partnership with leading conservation organizations in the field can make a difference by providing human, technical, and financial resources. Some organizations such as the Wildlife Conservation Society and the Frankfurt Zoological Society are recognized for their long-term commitments to vulnerable conservation hot spots throughout the world. Although zoos have successfully reintroduced some captive-bred species into the wild, e.g. the black-footed ferret, California Condor, Golden Lion Tamarin, etc., field partners recognize that ethical arks must not pursue counterproductive measures of reintroduction. For example, Luke et al., recently concluded that programs in Africa that offered contact with captive-born lions and an opportunity to help reintroduce them to the wild could not be regarded as a serious approach to species restoration. These programs are not operated or funded by accredited zoos and must be regarded as a significant departure from the norms of field conservation. Speaking on behalf of Panthera and its allies, Luke and his colleagues asserted:

> . . . approaches to reintroduction exemplified by the lion encounter industry do not address the reasons for the decline of lions *in situ*, nor do they represent a model that can be widely applied to restoration of threatened felids elsewhere. (p. 1)

As if photographic safaris were not exciting enough, wildlife entrepreneurs are offering encounters that are highly unusual. A tourist can ride a trained elephant to view wildlife populations in Africa. Since North America zoos of the AZA no longer approve of elephant rides, due to safety and animal welfare considerations, it is unlikely that eco-tours organized by ethical arks will patronize questionable animal encounters promoted as legitimate conservation experiences. As we have seen, conservation, ethics, and welfare are inextricably connected.

Affirming this conclusion, Paquet and Darimont (2010) also found no inherent conflict between conservation and animal welfare and suggested that the two organizational drivers are complementary. They acknowledged the importance of making this connection to counteract the anthropocentrism that has resulted in diminished welfare in wildlife populations throughout the world:

> . . . despite different conceptual underpinnings, advocates of animal conservation and animal welfare need to work toward a consistent and coherent ethical framework, with a willingness to recognize that no single value always or automatically trumps all other values . . . Individual animals, including humans, have value. Conservation has value. How we reconcile these values equitably is the ethical conundrum for all of us. (p. 187)

Chapter 2
Defining Animal Welfare

Animal welfare must not become irrelevant. Our task is to establish a workable, morally and scientifically acceptable way of dealing with the substance and the perception of paradox in our relations with wild creatures.

W.G. Conway

One of the most critical aspects of ensuring acceptable standards of welfare in zoos and aquariums is identifying and continually refining the most appropriate welfare measurement techniques. Without highly reliable measurement, welfare assessment relies on anecdotal and sometimes anthropomorphic judgments that may not accurately capture or reflect an animal's state of well-being. It is critical that zoos instead develop objective, reliable, and replicable measures that allow different individuals, at different time points in different settings or contexts, to assess welfare in a consistent manner. There is much to be learned from the welfare metrics of related fields, such as biomedicine or agriculture, but zoos also present a unique set of challenges such as the variety of species, low sample size and the housing of animals that have not been domesticated. Nonetheless, researchers have sought to overcome these issues and developed and used a variety of qualitative and quantitative measures to analyze welfare in zoo animals. The onus remains on zoo investigators to continuously validate and improve these measures, while finding ways to apply them to new species and situations.

Measurement is not a straightforward task as many factors influence zoo animal welfare. One major challenge is actually defining and conceptualizing welfare. Broadly speaking, the metrics of animal welfare have evolved in conjunction with the changing concepts and philosophies of animal welfare. Early conceptions of welfare targeted the prevention of suffering, thus many early measurements of welfare were aimed primarily towards reducing or eliminating negative behaviors, such as stereotypic behavior, or reducing associated physiological indicators, such as decreasing cortisol levels. More recently, scientists have begun to focus on identifying measures that indicate an increase in positive welfare (Yeates and Main 2008), and there has been a consequent shift in

DOI 10.1007/978-3-642-35955-2_2, © Springer-Verlag Berlin Heidelberg 2013

Fig. 2.1 Atlanta's menagerie at the turn of the century, was designed for intrusive public viewing (Zoo Atlanta Archives)

measures focusing on the standard of "good welfare." Another major challenge is that zoological institutions must ensure all aspects of animal welfare, including basic health, psychological well-being, and simulating natural living conditions (Fraser 2009). Sometimes the advancement of one of these aspects may hinder others, and zoo staff must identify the optimal balance. In this chapter, we will discuss some of the challenging theoretical considerations that must be addressed when developing a welfare assessment program, including how to define welfare, and identifying suitable comparisons from which to judge a zoo animal's welfare (Fig. 2.1). We remind the reader that zoos and aquariums share a common perspective on animal welfare, and the science of animal welfare applies to both institutions. In this book we use "zoo" as a generic term encompassing an entire spectrum of zoological institutions. When we mean to examine an issue specific to one or the other, we will single it out for emphasis. We have worked primarily in zoos, but we have a keen interest in aquariums. Until someone writes "Aquarium Animal Welfare" we trust that our findings can be applied usefully to the zoo and the aquarium.

There are a multitude of techniques that can potentially be used to measure welfare. However, it is a complicated issue and a myriad of challenges face researchers attempting to measure welfare. Given the difficulty of even defining what constitutes acceptable welfare, it is not surprising that selecting an appropriate measurement technique is daunting. Nonetheless, we must continue to refine and improve methods for welfare measurement. As a starting point, Mason and Veasey (2010) provide a good summary of what constitutes a useful measure of welfare:

> The best animal welfare research controls for known potential confounds (e.g. activity levels, time of day, or stage of estrus cycle); uses indices that are well-validated, and whose strengths and weaknesses are well-understood; tests clear hypotheses, selecting measures best suited for the specific questions under study; and typically uses multiple approaches, because, as we have

> seen, no one single welfare index is perfect. It is worth noting that poor quality welfare research, in contrast, fails to control for or acknowledge confounds or alternative explanations for findings; uses poorly chosen or validated measures, to test unclear hypotheses (an approach potentially fraught with circular reasoning); and often relies on just a single measure. (p. 250)

This description summarizes the most basic and important aspects of measuring welfare and provides a starting point for a discussion of welfare metrics, but there will often be institution-specific needs that factor into this process. Here we will briefly review some of the more specific, highly significant challenges regarding the measurement of animal welfare in the zoo and recommend directions for future research.

2.1 The Scope of Welfare

Early approaches to defining welfare focused on the elimination or reduction of negative states. The Brambell Committee (1965) released a document on farm animal welfare in the United Kingdom offering a definition of suffering that included discomfort, stress, and pain (p. 61). The committee advised that efforts to ensure welfare should "make it an offence to cause, or permit to continue, avoidable suffering so defined." These ideas were more formalized in a press release from the UK Farm Animal Welfare Council in 1979. This document encouraged that the following standards be met for livestock:

1. Freedom from thirst, hunger or malnutrition;
2. Appropriate comfort and shelter;
3. Prevention, or rapid diagnosis and treatment, of injury and disease;
4. Freedom to display most normal patterns of behavior;
5. Freedom from fear.

These principles came to be known as the Five Freedoms and were pivotal in the advancement of animal welfare worldwide. Current zoo welfare research still places a premium on reducing negative behaviors that might relate to the stress induced by violations of the Five Freedoms. As discussed in a paper by Swaisgood (2007), constant, uncontrollable stress is associated with negative welfare. Negative behaviors associate with stress, such as stereotypic behavior, or negative physiological indicators, such as evidence of Hypothalamic–pituitary–adrenal (HPA) activity, are common measures of animal welfare. The ideas encompassed by the Five Freedoms serve as a foundation of critical standards that must be met, but more recent conceptions of welfare have expanded beyond the focus on avoiding suffering and minimizing distress. We want to underscore that this does not diminish the critical necessity that the five freedoms be met as part of the effort to ensure zoo animal welfare, but more that the focus on reducing or eliminating negative welfare is an incomplete approach to the problem (Yeates and Main 2008) and further advances are necessary.

Most notably, the focus is shifting towards developing positive measures of welfare. Historically, the "absence" of negative welfare measures was considered an indicator of good welfare (Swaisgood 2007), but recent efforts have focused on

Fig. 2.2 Hard architecture makes zoo visitors feel sorry for animals; c. 1976 (T. Maple)

actually identifying specific instances of good welfare. This shift to focusing on what an animal "likes" or "wants" will require scientific rigor as the concepts are developed and defined to avoid an anthropomorphized view of positive welfare. Yeates and Main (2008) review this growing area of research, but several brief possible examples include facial expressions, vocalizations or play behavior that might indicate positive affect.

Increasing positive measures will benefit welfare in that the goal of achieving optimal states will eventually replace the goal of avoiding sub-optimal states. Furthermore, there may be important practical reasons for focusing more on positive welfare. For example, Yeates and Main (2008) discuss several reasons why incorporating positive measures into welfare policy is important:

1. societal value;
2. increasing caregiver welfare in case of human-animal bonds;
3. increasing positive measures often decreases negative ones;
4. rewarding good outcomes may be more effective than punishing negative measures;
5. more flexibility in policy-making.

Given these practical reasons, as well as the benefit to animals, it will not be surprising if the focus continuously shifts in this direction (Fig. 2.2). Of course, while positive measures should not necessarily take priority over the avoidance of negative factors such as fear and stress (Yeates and Main 2008), the inclusion of positive welfare can benefit the overall pursuit of animal welfare. Welfare can be seen as a continuum, not just bad to good, but good to great. For example, Koene and Duncan (2001) recommended zoos aim to provide a "luxurious" life, not one in which they merely survive. The application of this model encourages captive animals to thrive, but we must remember that a good life in the wild is a hard life, so luxury shouldn't lead us to abandon our principles of contingent work. Depending on how one defines luxury, it could be a pathway to obesity and inactivity. In discussing these ideas, Melfi (2009) laments the lack of information on many species due to a well-documented taxon bias in the species selected for study by zoo biologists. For example, this avian-mammalian bias leads to a focus on visual perceptions of the physical environment while olfactory, auditory, and electromagnetic sensory channels are ignored.

Barnett and Hemsworth (2009) have recently proposed several principles that are necessary to ensure animal welfare. These standards overlap the differing approaches to welfare and include the same provisions established in the five freedoms. These concepts promote several principles to safeguard animal welfare encompassed by the following assertions:

1. Minimize stress;
2. Minimize negative emotions;
3. Maximize positive emotions;
4. Ensure adaptation;
5. Provide opportunity for normal or natural behaviors; and
6. Provide natural environments.

By adopting frameworks such as this one—that include the avoidance or minimization of negative components *and* the maximization of positive components—zoo researchers will be best equipped to tackle the complicated issue of defining welfare for zoo animals.

2.2 Identifying the Focus of Welfare

Although it might seem like anyone interested in animal welfare would have the same goals and focus, this is not always the case. Fraser (2009) described three general categories of welfare focus: (1) basic health and functioning, (2) affective states, and (3) natural living approaches. The metric of choice in measuring welfare differs significantly depending on the focus, and can lead to differing views of how welfare should be measured (Barnett and Hemsworth 2009). For example, if the main concern is basic health and functioning, pathological and epidemiological measures may be most relevant. If affective state is the primary interest, communication signals (e.g. distress vocalizations) or fear responses could be recorded and used for analysis. Natural behaviors can be determined by observing wild counterparts, promoted by designing functionally appropriate environments for captive animals, and measured using standard behavioral data collection techniques based on an ethogram (Fraser 2009).

Ideally, what would be considered an improvement in one approach to animal welfare will result in improvements from other perspectives. For example, allowing an African warthog to wallow in mud will improve (1) basic health and functioning because body processes will not be disrupted by heat, (2) affective state because the animal will feel more comfortable, and (3) natural living because the warthog can perform a natural, thermoregulatory behavior. However, it is not always the case that improvements to one aspect of welfare improve others. These various philosophies

Fig. 2.3 Hard, sterile cages are designed for cleaning, not for living (T. Maple)

towards animal welfare have shaped different, and sometimes conflicting, approaches and assessments (Fraser 2009). For example, housing an animal in an isolated, sterile environment will improve basic health and functioning by reducing disease, but will reduce welfare for a social species if the natural living conditions are not met. This was the justification for maintenance-friendly facilities in the era of hard zoo architecture; sterility controlled the growth of bacteria and prolonged life in a highly unnatural setting (Fig. 2.3).

Not surprisingly, the philosophy an individual or organization adopts towards animals and their welfare will greatly influence what should be measured and how it should be measured. Because improving one aspect of welfare will sometimes impede others, it is important to determine what focus is most important when measuring welfare (Fraser 2009). Some organizations or individuals may only be concerned with one focus of welfare. For example, a veterinary hospital may house an individual from a social species in an isolated, sterile environment to reduce disease transmission rates and maintain basic health, but this impedes natural social behaviors and may increase distress related behaviors or contribute to a physical breakdown. Nonetheless, these decisions may be appropriate based on the responsibilities of the medical unit. However, we contend that zoos must be concerned with all three aspects of welfare—basic health and functioning, psychological or affective well-being, and simulating natural living conditions—which can be a challenging proposition since attempts to improve welfare may positively influence some measures while impairing others. For example, introducing zoo animals to live in a social group may temporarily increase aggression. This practice may have short-term negative effects on basic health and/or affect, but will result in long-term benefits in all three areas. Zoo researchers and managers must develop welfare protocols that optimize, to the extent possible, all aspects in the long-term. There are certainly common elements between the different approaches to welfare (Barnett and Hemsworth 2009), and by focusing on these similarities, we can identify the best measures to improve all aspects of welfare. As the field of zoo animal welfare progresses, it is particularly important for zoo researchers to work towards a convergent philosophy and measurement system for animal welfare.

There are other complications that can arise when encouraging natural behavior and naturalistic relationships. For example, in the famed Zurich Zoo Africa House, developed by Heini Hediger, the priority was to provide a mixed species experience. Among other birds, red-billed oxpeckers (*Buphagus erythrorhynchus*) were given access to black rhinos (*Diceros bicornis*) much as they would co-exist on the African savanna. Little is known about the relationship of oxpeckers to rhinos but it has been assumed that the birds help to remove external parasites. However, in captivity at the Zurich Zoo, rhinos have no parasites and the birds are well provisioned. The authors of a report evaluating this relationship, McElligott and colleagues (2004), discovered that red-billed oxpeckers were actually producing wounds on the rhinos and consuming blood. The rhinos were not tolerant of this behavior. Although the wounds were small, the phenomenon is an example of a naturalistic simulation that takes an unexpected turn. The birds have continued to live with black rhinos but the authors caution others to practice vigilant monitoring to ensure that the wound-producing habits of oxpeckers don't injure their hosts.

Barber (2009) recommended the wider use of species-specific Animal Care Manuals, currently being developed at a fast pace. At the time of his publication there were 160 manuals in production by AZA Taxon Advisory Groups. Once we have a standard for a given species, it is possible to utilize a "green flag-red flag" approach in identifying achievements or deficiencies for the institution in question. Welfare "red flags" would project a baseline level of concerns for a given population, and this would trigger careful monitoring over time. Green flag-red flag evaluations are a way to measure the effects of manipulations designed to improve welfare; just one more way to determine what works.

2.3 Influences on Welfare

Along the same lines as the focus of welfare, there are many different theories regarding the primary influence(s) on welfare. To progress as a science, it is critical to identify effective theories and rule out those that are obsolete or redundant. Swaisgood (2007) thoroughly describes ten such theories, although he does not advocate the validity of any one of them. He also provides a framework for minimizing or combining some of these theories, eliminating others and validating some. This process involves condensing redundant theories-identifying critical predictions and hypotheses-especially when these can be used to differentially support one theory, and determining the appropriate level of analysis. We must then review existing literature, or conduct new studies, to provide differential support for some ideas over others. We will briefly describe each of the theories discussed in his article, but refer the reader to the original work for a thoughtful and comprehensive review including citations to the primary sources of information for these ideas (Swaisgood 2007).

1. Ethological needs. Animals need to perform certain behaviors. In addition to the reinforcement of food or mating, animals are actually reinforced by the behaviors necessary to acquire these outcomes. So, for example, an animal that has evolved to dig food from underground has an inherent need to do so, and the ability to perform these species-specific behaviors is reinforcing and critical for optimal welfare. In a zoo setting, animals are often directly given food items, without requiring any manipulation or effort, and this could lead to sub-optimal welfare.
2. Information Primacy. Animals have evolved to seek out information about the location and availability of food and other resources. Similar to the ethological needs approach, information primacy suggests that animals have an inherent need to seek information by exploring the environment. The fact that animals have sometimes been observed to contrafreeload (i.e., work for food rather than get it for free) supports the information primacy concept. However, as previously mentioned, zoological husbandry procedures do not often require animals to seek out or manipulate food (sometimes enrichment does require this, but typically the majority of the diet is delivered prepared and ready for consumption). This might be a factor contributing to welfare in a zoological setting.
3. Mimicking Nature. A long-held belief within the zoological community is that nature and an animal's natural environment provide critical information about the best conditions for animals in captivity. Although there are some drawbacks to this approach that we will discuss in more detail later in this chapter, the natural environment serves as a critical yardstick upon which to measure some aspects of captive environments and can provide motivation and inspiration for developing new ideas. The extent to which captive environments differ from nature may be related to welfare issues. However, as will be discussed later, many aspects of the natural environment, such as natural disasters, starvation and drought, would most certainly yield bad welfare outcomes for animals in captivity, so this is far from being a straightforward issue.
4. Control/Choice. There is a large body of literature devoted to the concept of increasing choice and control over one's environment as a means to improve the welfare of animals living in captivity. This research has been carried out broadly in laboratory, farm and zoo animals with the broad consensus being that choice improves welfare. Given standard protocols in zoological environments, in which animals are typically provided with food or other resources independently from their own behavior or any effort, it can be seen how a lack of control/choice might lead to welfare issues. As an example, Jenny and Schmid (2002) provided tigers with electronically locked feeding boxes that would periodically open. Thus, continued "exploration" of the space was reinforced with the delivery of food, rather than the food being put in the exhibit irrespective of the animal's actual behavior.
5. Boredom. This idea suggests that animals in captivity lack sufficient behavioral stimulation and opportunities. Stereotypic behavior may develop in place of the natural behaviors that are not available to develop. Thus stereotypies may serve as a form of coping in the absence of other behavioral opportunities.

6. Lack of Sensory Stimulation. The theory regarding lack of sensory stimulation and welfare is closely related to the idea of boredom. More specifically, the idea is that there is not enough complexity or variability in the sensory information available in captive environments to properly stimulate an organism. The sights, smells, sounds, feelings and tastes available in a standard holding area are not only less variable than an animal might be exposed to in the wild, but may also represent sensory information that is not appropriate for an animal such as the smells associated with cleaning materials. Efforts to increase the variability and appropriateness of sensory information available in the environment may help to encourage natural behaviors and reduce undesired ones.
7. Stress. Stress itself is not a completely negative concept. It allows an animal to respond and adapt to changes in the environment and avoid or handle stressors. An animal has three types of biological response geared towards maintaining homeostasis when experiencing a threatening stimulus: behavioral, autonomic and neuroendocrine responses (Moberg 1985). The welfare issue arises if the stress response does not alleviate the stressor. For example, if a wild animal encountered a predator, a stress response would occur with physiological and behavioral changes that may allow the animal to escape. Once out of danger, the stress response would subside. However, in captive environments, an animal might have a stress response to a variety of unavoidable or unescapable stressors, such as large crowds. In this situation, the stress would become chronic and unpredictable, and a potential threat to welfare. However, as discussed in Swaisgood (2007), stress may be an intervening variable, not in and of itself a threat to welfare, but an event that occurs as a result of something else. Thus, the explanatory power of stress as a threat to welfare may be limited.
8. Coping. In a related theory, coping may be a critical function of stereotypic behavior and providing opportunities for adequate coping in animals may be necessary to permit the best welfare. This idea relies on the concept that an animal has certain mechanisms to alleviate or cope with stressors in the environment. When these opportunities don't exist, or are limited, an animal's welfare may be at risk. Some have argued that animals that engage in stereotypic behavior as a means of coping may actually have better welfare than those individuals who do not have any coping strategy. It is clearly a complicated issue and much more research is necessary to unravel this complex picture.
9. Behavioral channeling. According to this idea, in the absence of a variety of options, behavior may begin to take on very simple forms. This subset of behaviors, from the individual's overall repertoire, may become highly repetitive and occur at inappropriate times and/or contexts.
10. Perseveration. The psychological term, perseveration, refers to the continuation and repetition of a response that is no longer appropriate. Research suggests that perseveration and stereotypic behavior may be correlated, but there is still a great deal of research required to tease out the causality of this relationship.

As Swaisgood (2007) discussed, this is by no means an exhaustive list of theories, but highlights some of the main ideas in the literature. He also carefully describes the process by which alternatives can be ruled out or supported. This is a critical step moving forward so that zoo welfare research is theoretically driven and contributing to a larger, broader picture.

2.4 Identifying Appropriate Comparisons

Another challenge in measuring welfare is determining an appropriate source for comparison. Welfare can only be considered "good" or "bad" in relation to something else (i.e., across individuals or within individuals at different times or in different contexts, etc.). For example, one could compare an individual's behavior in various conditions, or make comparisons across groups in different conditions. A commonly referenced comparison for zoo animals is the wild or natural environment, although the issue is complicated. Many behavioral and environmental characteristics that are similar to those found in the wild are desirable (Maple and Bloomstrand 1988), and unnatural, stereotypic behaviors are not ideal. Thus, natural behaviors should be an important consideration for many aspects of zoo management, but replicating the wild environment may not always lead to optimal animal welfare. In the wild, animals face a variety of threats, such as famine, drought, and predation (Hutchins 2006), and exhibit a variety of behavioral responses such as shivering from cold or fleeing in panic from predators (Fraser 2009). Although these conditions and behaviors are "natural", they are not desirable for zoo animals. Exposing animals to these natural conditions would threaten animal welfare, and replicating nature may be undesirable in this context.

Fraser (2009) offers one potential solution to the problem of replicating natural conditions, but maintaining optimal welfare. By incorporating natural conditions and animal preferences, we can identify the optimal conditions for zoo animals. As depicted in Table 2.1, this concept can be represented by a 2 × 2 matrix, with the most ideal condition for zoo animal welfare being the natural/preferred condition.

Providing unnatural/preferred or natural/not preferred conditions may not ensure optimal animal welfare, and may even be detrimental (Fraser 2009). For example, although sweetened foods may be preferred, overconsumption of sugar may lead to obesity problems in captive animals. Therefore, we argue that an approach such as the one described by Fraser will be best for determining which aspects of the environment are both naturalistic and preferred. This technique is also consistent with Dawkins (2008) approach to measuring welfare that asks (1) are the animals healthy? and (2) do the animals have what they want?

2.5 Species-Appropriate Measures of Welfare

Identifying an optimal or ideal level of various environmental inputs is another challenging, yet important component of welfare. It is not simply the case that maximizing any aspect of husbandry that is considered good for welfare will benefit the animal. For example, although increasing novelty in the environment is beneficial,

Table 2.1 Adapted from Fraser (2009)

	Natural	Unnatural
Preferred by the Animal	**Natural & Preferred** e.g. warthog wallowing in mud	**Unnatural & Preferred** e.g. eating sweetened foods
Not Preferred by the Animal	**Natural & Not Preferred** e.g. fleeing from predator	**Unnatural & Not Preferred** e.g. avoiding water hose

maximizing novelty will reduce control, which is also an important component of welfare (Yeates and Main 2008). Thus, a continuum may be the most appropriate representation of welfare, with the ideal value sometimes occurring at the midpoint. It is likely that these values will vary across species; thus, it is important for zoo researchers to publish and present findings in order to avoid unnecessary replication.

Whitham and Wielebnowski (2009) advocated a three step process for developing a species-specific monitoring tool for individual animal welfare. First, a welfare score sheet is developed based on animal care experts (keepers/managers) responses to a detailed welfare assessment questionnaire. Second, these welfare score sheets are validated by determining if the scores are correlated with behavioral and physiological measures over a 6 month time period. Finally, these welfare score sheets can be used by animal managers to proactively address changes in well-being. The Chicago Zoological Society has started this three step process for a variety of species: aardvark, African elephant, black rhino, clouded leopard, fennec fox, Goeldi's monkey, green-winged macaw, leopard gecko, okapi, polar bear, red-tailed hawk and western lowland gorillas (Whitham and Wielebnowski 2009). If these welfare score sheets can be validated and are found to be reliable, this will be an important advancement in the assessment of individual animal welfare. Zoos can all contribute to developing score sheets for different species, and once this technique is considered valid and reliable, welfare score sheets can be shared among zoos. This will provide a low-cost, effective technique for monitoring and evaluating zoo animal welfare. A recent international survey of managers, keepers, educators, veterinarians, scientists, and animal welfare experts who rated variables associated with great ape husbandry provided additional evidence that experts can be used to evaluate welfare (Fernie et al. 2012). We will review the results of this survey in the context of zoo design in Chap. 8.

2.6 Need for Multiple Measures

Given the wide variety of tasks used to measure welfare, and the many complexities and challenges of studying welfare, it is critical to avoid reliance on a single measure, but include multiple measures when analyzing welfare. Behavioral and physiological measures are sometimes, but not always, correlated (Barnett and Hemsworth 2009). Also, because similar autonomic states may reflect drastically different "emotional states," Dawkins (2008) recommends the possible solution of incorporating reinforcement theory into welfare measurement. She describes an example of riding a roller coaster. Individuals that enjoy the ride and individuals that are terrified by the experience

Fig. 2.4 Wild mountain gorillas, the prototype of soft architecture (J. Fowler)

will exhibit similar responses (e.g. screaming, white knuckles, increased cortisol). The only way to determine if the experience was positively reinforcing is to see if the person rides the roller coaster again (Dawkins 2008). Incorporating a measurement of choice or preference will improve our understanding of what features of the environment improve zoo animal welfare.

There is also a pressing need for scientific research to validate metrics of welfare in zoos. This can be done by determining if different measures are correlated with one another (Dawkins 2008, Barnett and Hemsworth 2009). Some measures have been validated for domestic or production animals, but these may not be valid for zoo animals (Melfi 2009). As researchers refine and validate measures, it may be possible to reduce the number of measures needed, but in the current state of the zoo welfare literature, we suggest that it is important to incorporate as many measures as possible. Researchers should also be encouraged to publish information on measures that could not be validated with established measures of welfare, as the dissemination of this knowledge will aid other institutions and help advance the field.

Another aspect to consider when selecting measures is whether the focus will be on positive or negative aspects of welfare. As discussed previously, historically, perhaps resulting from a welfare conception related to reducing and preventing suffering, studies focused on the reduction or absence of negative measures. For example, the absence or reduction of "negative" welfare indicators such as decreased stereotypic behavior or baseline cortisol levels are often used to indicate satisfactory welfare. However, some exotic species may not have evolved overt signals to convey unsatisfactory welfare, as expressions of weakness would be detrimental to a wild animal's survival (Melfi 2009). Thus, we must move away from the overreliance on the presence/absence of negative indicators, and zoo researchers need to also establish reliable and valid measures of positive welfare (Fig. 2.4). At this point in time, behavioral measures may be particularly useful for measuring positive welfare because reliable physiological indicators have not been established (Yeates and

Main 2008) although several possibilities exist and continued exploration of this area will likely yield exciting results. We suggest that when measuring the outcome of a welfare intervention, it is important to reduce negative measures *and* increase positive measures. Future research should focus on continuing the development of positive measures of welfare by refining behavioral techniques and identifying physiological markers.

Chapter 3
Welfare Metrics Applied

The possibility of stepping into a higher plane is quite real . . .
It requires no force or effort or sacrifice. It involves little
more than changing our ideas about what is possible.
Deepak Chopra

There are many significant challenges facing zoo welfare researchers. Despite the difficulties, many attempts to improve and measure animal welfare have been made. Broadly speaking, welfare can be measured as inputs (i.e., environmental characteristics) and/or outputs (i.e., animal characteristics). Inputs, such as space allowance, food and enrichment, provide consistent and objective measurements of the environment (Yeates and Main 2008). Zoo and aquarium specialists have a long history of modifying inputs such as exhibit design, space restrictions, feeding schedules, social housing conditions and enrichment in an effort to improve animal welfare (e.g. Erwin et al. 1979; Maple 1980a; Bashaw et al. 2003; Hosey 2005; Resende et al. 2009; Carlstead et al. 1999). Accordingly, zoo policies or regulations often focus on minimum requirements for inputs under the assumption that this will result in improved welfare (Whitham and Wielebnowski 2009). However, the provision of positive inputs does not necessarily ensure good welfare (Yeates and Main 2008) and does not address *individual* animals (Barber 2009); therefore, it is critical to also measure the *outcomes* of inputs. The outputs, or animal responses to welfare inputs, can be measured behaviorally and physiologically (Latham 2010; Barnett and Hemsworth 2009). This "toolbox" of measures can be used, most effectively in combination, to assess welfare (Latham 2010). Here we will discuss a variety of behavioral and physiological outcomes that can be measured at the zoo.

T.L. Maple and B.M. Perdue, *Zoo Animal Welfare*, Animal Welfare,
DOI 10.1007/978-3-642-35955-2_3, © Springer-Verlag Berlin Heidelberg 2013

3.1 Preference and Motivation Tests

An important goal for animal care staff is to provide animals with resources (e.g. food, environmental characteristics) that are *both* preferred and motivating. Providing an animal with preferred resources is an important component of ensuring welfare, but may not meet the goals of all welfare interventions. For example, preference for a resource does not ensure that an animal will be motivated to obtain it and may not result in behavioral change. Thus, it is important to note the difference in measuring preference and motivation and the implications of the findings from these different types of tasks. Choice tests and operant tasks are two broad, but not mutually exclusive, techniques used to address preference and/or motivation (Kirkden and Pajor 2006). Choice tests provide an animal with a simple choice between alternatives and are ideal for measuring preference. Operant tasks require a subject to perform a behavior (e.g., lever pressing) in order to gain a resource (e.g., food) and may be better suited for measuring motivation to obtain a particular resource. In zoos, both choice and operant tasks have been used to measure preference and motivation. Food choice tests have been used in a variety of zoo animals (e.g., colobus monkeys, Tovar et al. 2005; fruit bats, Masefield 1999; giant pandas and African elephants, Gaalema et al. 2011) in an effort to determine what resources are preferred. Choice tests have also been used to measure preference for other features of the environment, such as preferred floor substrate in elephants (Meller et al. 2007). One interesting possibility is to provide cognitively-advanced animals the opportunity to select a resource from a presentation of photographs. A study of western lowland gorillas found that preferences for photographs of food were consistent with their preferences for actual food items (Clay et al. 2009). This technique may be particularly useful in providing zoo animals with a choice between large enrichment items or access to various enclosures that would otherwise be impossible to present. The vast cognitive repertoire of cetaceans provides an opportunity to examine their environmental preferences and may shed light on welfare as evaluated by dolphins and whales themselves (e.g. Herman 2010).

Although providing an animal with a preferred resource is an important component of ensuring welfare, it does not guarantee improvements in welfare because animals may not be motivated to obtain a certain resource, even if it is preferred when presented with a variety of alternatives. Operant tasks are better suited to determine which resources are the most motivating or reinforcing (i.e., increase rates of behavior). The use of optimal reinforcers will improve the effectiveness of enrichment or training programs. One zoo study found that highly preferred items served as effective reinforcers on an operant task in both elephants and giant pandas (Gaalema et al. 2011), however, it is important to empirically identify effective reinforcers if the goal is to increase rates of behavior. Another particularly important finding of that study is that keeper ratings of preference did not always correlate with the animal's actual preference, suggesting the need for systematic evaluation and highlighting the need to have scientific personnel on zoo staff to conduct thorough, unbiased scientific evaluations of animal welfare.

Measuring what an animal prefers or selects is a valuable tool for improving welfare, but there are some recognized problems inherent in preference testing. As discussed in Yeates and Main (2008), context may alter preferences. For example, animals are only motivated to perform certain behaviors once prepotent, or the most pressing needs such as thirst or hunger, are satisfied. Preference assessments only provide a relative value judgment compared to other test items, not an absolute value. Futhermore, animals often prefer options that benefit short-term welfare, but not necessarily long-term welfare. For example, an animal may prefer and be highly motivated to obtain foods with high sugar content, but this could have long-term health consequences that do not optimize welfare in the long run. It is important to assess the long-term effectiveness of changes made based on preference assessments (Latham 2010).

3.2 Patterns of Behavior

It has also been proposed that the organization or pattern of behavior may be a critical measurement tool, with increased variability or flexibility generally thought to reflect improved welfare. Asher et al. (2009) discusses several lines of evidence that support this relationship between variability and welfare. Greater environmental complexity leads to greater neural complexity which creates the potential for more behavioral variability. If behavior is more variable, more behavioral options are available which result in increased control over the environment. Also, stress decreases physiological variability and likely has the same effect on behavioral variability. Thus, decreased behavioral variability may reflect deficits in welfare, and increased variability may reflect improved welfare. There are several analytical techniques such as fractal analysis, temporal analysis, social network analysis and agent-based modeling, which can be used to measure behavioral organization and potentially improve animal welfare (for a review see Asher et al. 2009).

Zoo researchers may also examine variability in the use of space or resources (e.g., Ogden et al. 1993; Ross et al. 2009; Forthman-Quick and Pappas 1986; Bettinger et al. 1994; Horikoshi-Beckett and Schulte 2006; Leighty et al. 2010) as an indicator of good welfare. For example, Ross and colleagues (2009) compared the use of space in an older exhibit and a newer exhibit based on preferences observed in the older location. Electivity indices were used to compare gorilla and chimpanzee use of environmental features to their availability in the environment, which can be used as a consistent measure of space use and animal welfare (Ross et al. 2009).

3.3 Natural Behavior

An aim of many welfare assessments is to increase natural behaviors. As will be discussed in detail below, there are some caveats with this approach, and not all natural behaviors (e.g., shivering or fear responses) indicate improved welfare. Nonetheless, zoos often design enclosures and enrichment programs in order to increase natural behaviors. For example, Kawata (2008) emphasizes the importance of incorporating species' natural feeding conditions into the design of feeding protocols in the zoo. In addition to providing food with appropriate nutritional content, zoos should consider a variety of factors such as frequency, predictability, seasonality, and mode of food presentation (Kawata 2008; see also Maple and Bloomstrand 1988). As Keeling et al. (2011) observed, natural behavior that enhances the behavior of one individual can, at the same time, lead to a reduction in welfare for another member of the group. In her review, she offered a definition of (positive) natural behavior according to Bracke and Hopster (2006):

> Natural behavior may be defined as behavior that animals have a tendency to exhibit under natural conditions, because these behaviors are pleasurable and promote biological functioning.

It is also possible to evaluate an animal's welfare state using a species' natural auditory, visual, or olfactory signals. Species-specific signals have evolved to convey information, such as pain, distress, or pleasure to conspecifics. Zoo researchers can use these signals to gather information on an animal's state of welfare. For example, research suggests that elephant rumble vocalizations may reflect affect (Soltis et al. 2009), and as this field of research develops, these vocalizations can potentially be used to measure good or bad welfare (Mason and Veasey 2010). However, zoo researchers must avoid anthropomorphizing behaviors (e.g., interpreting liquid running from the eyes as "crying" with emotional meaning) and be aware that not all signals are perceptible to humans (e.g., vocalizations out of range or olfactory signals) (Mason and Veasey 2010).

3.4 Cognitive Bias Assessments

Recently, a technique known as cognitive bias has been introduced into the toolbox of welfare measurement. This technique is derived from human research which finds that people in a negative affective state (e.g., depressed or in a negative mood) will perceive neutral stimuli as more negative than individuals in a positive affective state. Studies in rats and birds have implemented go/no-go and conditional discrimination tasks in an effort to measure cognitive bias. The subject is trained to respond in a particular manner to a positive or negative stimulus. After this training is completed, neutral stimuli are introduced. Responses that are consistent with the positive stimulus indicate a positive cognitive bias and responses that are consistent with the negative stimulus indicate a negative cognitive bias. Changes in environmental or biological conditions have been found to induce the expected shifts in biases (either more positive or more negative).

Rodent researchers have found that a short-term manipulation of emotional state results in a decreased response to ambiguous stimuli for subjects that changed from low-to-high anxiety conditions, and an increased response from subjects changed from high-to-low anxiety conditions (Burman et al. 2009). A recent study found that a positive change in environmental conditions can actually induce an "optimistic" cognitive bias where subjects were more likely to interpret ambiguous stimuli as an

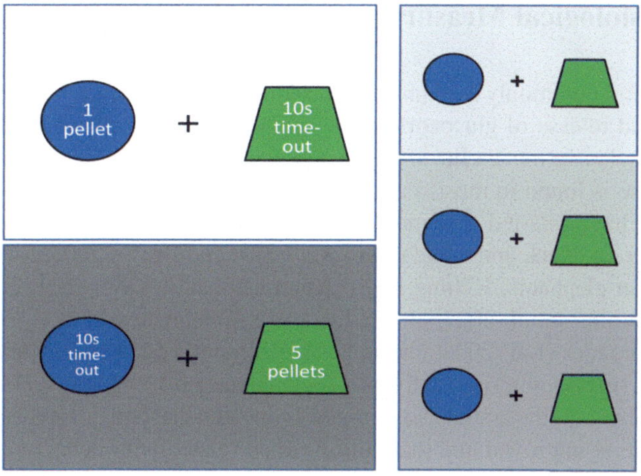

Fig. 3.1 An example of a cognitive bias task in which subjects are first taught a conditional discrimination (*white background*, touch the *blue icon*; *gray background*, touch the *green icon*). Then, intermediate levels are introduced (*various shades of gray* shown in *second column*) to determine if subject responds as if the background is *light* (less positive) or *dark* (more positive)

indication that a higher value reward was available (Brydges et al. 2011). Cognitive bias may be a useful technique to analyze welfare states of zoo animals, especially considering the potential to measure both negative and positive changes in affect (Fig. 3.1).

3.5 Stereotypic Behavior

Stereotypic behavior, or "repetitive, invariant behavior patterns with no obvious goal or function" (Mason 1991) is one of the most commonly used measures of compromised animal welfare (Mason and Veasey 2010). Stereotypic behavior is discussed in depth elsewhere in this book, but it should be acknowledged that reducing or eliminating stereotypic behavior is associated with improved welfare and is one of the best validated measures that we currently have of animal welfare. In addition to a large number of animal studies (e.g. Holzapfel 1938; Berkson and Mason 1964; Sargent and Keiper 1967; Erwin 1979) stereotyped behavior has also been extensively documented in institutionalized human beings (Hutt and Hutt 1965; Kaufman and Levitt 1965; Baumeister and Forehand 1973).

3.6 Physiological Measures

One of the most commonly used physiological measures of welfare is the activation of HPA axis and release of glucocortoicoids, as indicated by corticosteroid or cortisol levels. Cortisol is the metabolite of glucocorticoids found in most large mammals, and corticosterone is found in most small mammals and birds (Wasser et al. 2000). Zoo researchers have started to incorporate cortisol measurements into welfare assessments (e.g. black and white rhinos, Carlstead and Brown 2005; Linklater et al. 2009; African elephants, Kelling 2008; Asian elephants, Laws et al. 2007; *maned wolves*, Cummings et al. 2007; clouded leopards, Wielebnowski et al. 2002a; snow leopards, Burgener et al. 2008; cheetahs, Wielebnowski et al. 2002b; giant pandas, Snyder et al. (in review); spotted hyenas, Goymann et al. 1999; tigers, Dembiec et al. 2004). The results of these studies are mixed. Some find that cortisol levels are reduced when welfare is improved and that cortisol is correlated with behavioral measures, however, others find no relationship between cortisol levels and improved welfare. Thus, it is critical to continually revise our understanding of the relationship between cortisol, behavior, and welfare. It is also important to avoid using cortisol as the only measurement, as it can be confounded by many other variables, and there are many difficulties with the process of measurement that need to be recognized when designing a study and selecting data collection techniques.

As summarized by Latham (2010), there are pros and cons to each glucocorticoid sampling method. Blood and saliva samples provide an accurate measure of acute stress levels, but only measure one time point, and may be affected by circadian rhythm. Blood samples are highly invasive and may be difficult to obtain, particularly without introducing additional stress and potentially increased HPA axis activation as a result of the sampling technique itself. Although saliva sampling is less invasive than blood sampling, it requires a great deal of training and may not always be successful. Fecal/urine and hair sampling are both non-invasive which may make them more useful for zoo researchers, however, these forms of sampling are not without limitations. Hair samples are not susceptible to circadian rhythm, but only provide a measure of chronic stress. Fecal/urine samples measure the average level of glucocorticoid over several hours, and multiple samples need to be collected to obtain acute stress measures. It is critical to collect samples in the most appropriate ways to address the welfare question of interest. Zoo researchers need to be aware of the limitations of each of these methods in order to draw appropriate conclusions from the data. Furthermore, HPA axis activation is not limited to aversive stimuli, and may be activated during other arousing situations (Latham 2010). This is a particularly challenging situation for researchers and highlights the need to measure additional behavioral or cognitive factors to validate that increased cortisol levels relate to negative states. Research that relies on single measures of HPA activation may severely overestimate *or* underestimate the implications of the findings for animal welfare (Fig. 3.2).

There are a variety of other hormones that may provide useful information about welfare. Of particular interest is the possibility that some hormones, such as oxytocin, may be used to measure positive welfare states (Yeates and Main 2008). Zoo researchers are in a good position to advance this field of study. Animals are frequently moved to new enclosures, or husbandry routines are changed in an effort to improve positive welfare. By systematically measuring hormone levels before and after a change, and in conjunction with other behavioral measures, we can begin to unravel the underlying hormonal changes associated with positive welfare.

There are other important physiological measures that should be included in an overall assessment of welfare. The autonomic nervous system consists of two complementary systems: the sympathetic and parasympathetic nervous system. As part of the autonomic stress response, the sympathetic nervous system prepares an individual for "fight or flight". The parasympathetic system has been referred to as the "rest and digest" system and prepares the body for the activities that take place during resting states such as urination, digestion, and sexual activity. We can measure activation of the sympathetic nervous system by monitoring for increased heart rate, respiration rate, pupil dilation, blood pressure, and adrenaline (epinephrine) and noradrenaline (norepinephrine) levels. On the other hand, activation of the parasympathetic system may result in increased defecation and urination. These various measures of the autonomic stress response are correlated with acute stress and fear (Mason and Veasey 2010), and can be used as an additional technique to assess welfare in zoo animals. However, it should be cautioned that these measures increase with general arousal, so may be elevated by activities such as copulating or playing (Mason and Veasey 2010). Thus, as we have described with other measures, it is essential to incorporate multiple behavioral and physiological measures when attempting to evaluate welfare in zoo animals. Nonetheless, as we have illustrated in this chapter, there are a number of validated and potentially useful techniques for measuring welfare at the zoo, and with continued scientific research we can continue to make progress.

The zoo community needs to continue developing its own rigorous techniques for measuring welfare, and we have pointed out some of the substantial challenges associated with this goal, and some potentially useful measures that are specific to zoos. However, researchers should also seek guidance from research in the domestic and laboratory animal welfare field. Although there are issues that are only relevant to zoos and aquariums, there are many commonalities regarding animal welfare, and collaborative work will continue to advance animal welfare in all domains. We have focused on the metrics of zoo animal welfare, but we encourage readers to explore other sources that investigate these issues in a diversity of settings and circumstances (Animal Welfare series, Springer publications).

3.7 Research Design

There are several factors that make research design particularly important when assessing welfare in zoos. First, the sample size for zoo and aquarium studies is often very small. Solitary species such as red pandas or black bears are typically housed alone, and may be the only member of the species at an institution (Fig. 3.3). Even social species, such as gorillas or meerkats, are often housed in groups of less

Fig. 3.2 Fighting over a food resource in a flock of flamingos. Fighting behavior can lead to increased hormone levels indicative of HPA activation (B. Perdue)

than five or ten animals given the logistical constraints of housing captive wildlife. Thus, designing appropriate studies is of critical importance given that typical research designs rely on much larger samples. Furthermore, it may be the case that only one individual exhibits a stereotypical behavior or sign of distress that needs to be addressed. Given this limitation in sample size, careful planning should be taken in designing appropriate studies. The experimental analysis of behavior (EAB), a field of research based on the pioneering work of B.F. Skinner also suffered from small sample size, but the precise conditions of their research resulted in little variability, demonstrating strong and reliable outcomes. Because the field of psychology didn't immediately accept the tightly controlled small N research of the behaviorist school, EAB organized their own journals to publish their research (*Journal of the Experimental Analysis of Behavior*; *Applied Behavior Analysis*). Small sample size can be overcome by exerting careful control over extraneous variables and by the design of elegant experiments. Zoo and aquarium research, however, must take into account the considerable variation due to individual differences.

When possible, designs that present several iterations of a new input compared to a baseline (i.e., repeated baseline measures) should be implemented. For example, if assessing the impact of a new exhibit, there should be multiple alternations between the old exhibit and the new exhibit. By doing this, the effects of novelty or any other unmeasured third variables can be controlled. The improvement in welfare, however it is being measured, should fluctuate in conjunction with the change in housing condition *if* that is the critical input. Also, when possible, only a single variable should be manipulated at a time to be able to specifically identify which inputs improve welfare. Zoo carestaff might introduce a variety of enrichment devices all at once to try to help an animal exhibiting stereotypic behavior. Although this approach may benefit the animals being studied, it does little to advance our overall understanding of which types of enrichment are actually effective (Swaisgood 2007). If trying to measure the impact of new enrichment devices, one should not incorporate multiple items into an animal's enclosure at one time, rather, items should be introduced individually and in repeated iterations as described above. Or, if the exhaustive stimulus approach is necessary to alleviate a severe welfare concern, it

Fig. 3.3 Solitary species, such as the red panda, pose a particular challenge to zoo researchers given the small population housed at any single institution (B. Perdue)

should be followed up with attempts to systematically remove and replace particular items in an effort to identify which item actually influenced behavior and improved welfare.

There are some potential limitations to this approach. In particular, if there is a significant welfare concern, introducing multiple enrichment devices at once may be the best approach given the importance of addressing the welfare issue. Similarly, if the "baseline" condition is suboptimal, it would not be ethical to repeatedly expose an animal to that situation. Rather, researchers should collect as much data as possible before the animal is moved to a new situation. Of course, this highlights the importance of proactive welfare assessment in zoos as opposed to reacting to problematic situations. If the proactive approach is taken, much more control, through proper research design, will be possible.

3.8 Data Collection

We will briefly describe several types of data collection techniques that may be useful to zoo and aquarium researchers, but it should be noted that these are covered in greater depth in volumes dedicated to the topic (Martin and Bateson 2007). The interested researcher should examine these sources for more detailed information on data collection techniques. For a more detailed discussion on sampling methods, the reader should refer to a comprehensive book chapter by Crockett (1996). We have included a table summarizing sampling methods in Table 3.1.

Recording all of an animal's behavior would not be an effective use of anyone's time, thus researchers rely on data collection techniques that allow us to get a sample of behavior that accurately reflects the larger behavioral pattern. The type of data

Table 3.1 Options for statistical tests in various situations

	Parametric test	Nonparametric test
Pre-post design Subject or subjects are exposed to two conditions. *Example*: *A baseline condition compared to an enrichment condition*	Independent samples *t*-test	Wilcoxon Signed-ranks test
Repeated-treatments design Subject or subjects are exposed to repeated alternations of at least two conditions *Example*: *Data are collected in a baseline condition, then with enrichment, then another baseline with no enrichment, and then another enrichment condition.*	Repeated measures analysis of variance (ANOVA)	Friedman's test
Between-subjects design Different subjects are exposed to different treatment conditions *Example*: *Comparing one group of animals living in a highly enriched enclosure to another group living in an unenriched condition*	Analysis of variance (ANOVA)	Kruskal–Wallis test

collection selected should be based on the interests of the investigators. *Ad libitum, or ad lib, sampling* involves taking notes on as much behavior as possible. This could be used in preliminary studies or field work, and is useful for identifying broad patterns of data, but is not necessarily reliable because an individual cannot keep track of all things at all times and will likely be biased towards certain events (e.g., recording information about two individuals fighting without keeping track of feeding behavior of other animals in the group). Several systematic approaches to data collection exist that control for experimenter bias. *Focal-animal sampling* refers to recording the activities of a specific individual for a pre-determined amount of time. This sampling technique would capture the duration of behavior as well as the occurrence of specific events. *All occurrence sampling* involves recording all instances of a particular behavior. For example, one might record information about all instances of fighting in a group. This method will yield the rate with which a behavior occurs. *Instantaneous or scan sampling* refers to the technique of recording behavior at a specified interval (e.g., every 30 s). This can be done with a group where the researcher "scans" the behavior of group members at the interval. This sampling technique provides a good method for collecting a large amount of data, especially for a group. Less frequent events, such as vocalizations, might not be captured by scan sampling alone. *One–zero sampling* refers to recording whether or not a behavior or event occurred at all during a specified interval (e.g., 30 s). For example, one could use one–zero sampling to record vocalizations. An interval in which a vocalization occurred would be scored as a one regardless of how many vocalizations actually occurred during the time period and one in which no vocalizations were recorded would be scored as a zero.

These different sampling techniques have pros and cons depending on the question of interest. They can be combined to meet researcher needs, but this should be planned and designated prior to the onset of data collection. One other important consideration when collecting data is that multiple observers need to undergo reliability checks. This is to ensure that the same behaviors will be recorded in the same manner by different people. A carefully defined ethogram, or description of behaviors, is one aspect of ensuring reliability in data collectors. Reliability can be checked by hand or many software packages now offer reliability check options.

3.9 Analyzing Measurement Data

Another essential consideration for zoo animal researchers is how to analyze data. This is an important decision regardless of which measure is selected. Here we will briefly describe several statistical approaches that can be used in welfare metrics.

3.9.1 Pre-Post Tests (Within-Subject Design)

For within subject tests, or test in which the same animal experiences multiple conditions, a pre-post or repeated treatments design can be used to measure the behavioral and physiological outcomes of welfare interventions or inputs. This can be done at the individual level, where an animal essentially serves as its own control, or at the group level, where a number of animals experience the same repeated conditions (although not necessarily in the same order). Depending on data characteristics (e.g., scales of measurement or sample size), there are parametric and nonparametric tests for these types of designs. For frequency data or categorical data in which there are two conditions (e.g., pre- and post-treatment), nonparametric tests such as the paired-samples Wilcoxon signed ranks test can be used, while a paired-samples *t*-test can be used for data meeting assumptions for parametric testing. Pre-post designs are convenient when an animal is moved to a new enclosure or an enrichment program is started, but this measure is easily confounded by other time-related factors such as seasonal changes (Swaisgood and Shepherdson 2005). Post-occupancy evaluations (Chap. 7) are examples of research that generally deploys this type of design (Maple and Finlay 1986).

3.9.2 Repeated-Treatment Design

A better approach is the repeated treatment design, in which treatments and baseline or control conditions are presented multiple times, in a randomly alternating order that is predetermined (and can be counterbalanced across individuals). Repeated-treatment designs can control for confounding variables and has higher internal validity (Swaisgood and Shepherdson 2005). Order effects that might influence pre-post designs can be controlled for with repeated treatments. For data meeting the assumptions of parametric tests (i.e., normality, homogeneity of variance, independent data points), a repeated-measures analysis of variance (ANOVA) could be used to analysis these data, while a Friedman's test provides a nonparametric alternative.

3.9.3 Between-Subjects Tests

Between-subject comparisons can be made if two (or more) groups are exposed to different conditions. Between-subjects tests would be useful if two groups of individuals were housed in different conditions or subjected to different management techniques. For example, if a new enrichment device was introduced to one family group, but not another, these two groups could be compared using an independent samples *t*-test (parametric) or a Mann–Whitney *U* test (nonparametric). If more than two groups were involved, an analysis of variance (ANOVA) would be appropriate, with a Kruskal–Wallis test serving as the nonparametric alternative.

3.9.4 Multi-institutional Studies

Zoo research design is often hindered by small sample size, and one way to overcome this issue is to perform multi-institutional studies or survey multiple institutions. For example, Bashaw et al. (2001) surveyed 49 institutions about stereotypic behavior in giraffe and okapi. This approach yielded a significantly larger study sample than would have been available at any one institution. Survey data is an extremely useful method for gathering a great deal of information, but it is important to validate survey findings with observational or experimental research. Perhaps the most effective technique for zoos to use is multi-institutional, repeated-treatment experimental design (Swaisgood and Shepherdson 2005). Although this will be challenging to implement, the connectivity of zoo personnel has continued to increase over the years through TAGs, SSPs and other advisory groups, and the implementation of these types of studies should be a priority for these groups, especially for species that are particularly prone to developing stereotypies or other indicators of negative welfare. Another useful tool for zoo researchers is a meta-analysis based on the existing literature. In a meta-analysis, the results of many different studies are compiled and analyzed to determine if there is an overall effect of a particular treatment or intervention. For example, meta-analyses have been used to evaluate the effectiveness of enrichment on reducing stereotypic behavior in zoo animals (Swaisgood and Shepherdson 2005; Shyne 2006). These types of analyses can be used on any of the behavioral and/or physiological outcome measures discussed in this chapter.

An important paper by Kuhar (2006) summarized the methodological issues inherent in zoo research and appealed for resolution and uniformity in the use of statistical tests. He specifically examined how authors in the journal *Zoo Biology* interpreted data and found deficiencies in many cases, although compared to other journals the authors were not out of line with conventional practices in other specialties in biology. What is required, according to Kuhar, is a more consistent set of guidelines on what is acceptable practice going forward. By requiring standardization for data analysis, the science of zoo biology will be significantly advanced. Kuhar specifically recommended complete disclosure of data manipulation, reporting of statistical values, control for institutional effects in statistical models, and avoidance of pseudoreplicated observations. In addition, he advised the use of hierarchical or factorial models and randomization tests as a way to move beyond standard *t*-tests and ANOVA. A greater interest in research design and statistics in zoo biology will help to resolve some of the discrepancies in management practices. The controversy over survival data in wild and captive elephants is a case in point. The research of Wiese (2004) and Clubb et al. (2008) differed in the way the data were analyzed, with a completely different conclusion reached by the respective authors. The fact that Clubb's study was published in *Science* strengthened its impact and provided fuel for the controversy. As the field of zoo biology matures, its methodological rigor will be reflected in a higher quality of research and more robust findings.

Another important consideration is that statistics should be presented as part of the picture of an animal's welfare, not relied on exclusively. Researchers should publish means and standard errors, even raw data if possible, rather than only reporting the output of statistical tests. Further, statistics should not be applied if the assumptions of the tests are not met. One should refer to statistical textbooks to determine the assumptions of each statistical test and whether a dataset violates those assumptions. Given the small sample size and lack of independence in data points often found in zoo datasets, there are frequently limits on which, if any, statistical tests are appropriate. As a community, if zoo researchers continue to run well designed studies, apply appropriate analytical techniques, and publish their results, we can continue to discover, test, and export useful measures and approaches to assessing animal welfare.

Recognizing the unique settings where zoo biologists practice their science and their craft, Crockett (1996) provided useful guidelines for conducting behavioral research:

1. Formulate a specific research question.
2. Keep data collection simple.
3. Perform preliminary analyses on some sample data before finalizing the data collection design.
4. Collate and begin to analyze data while data collection is in process.
5. If the results of the study seem to be of general interest, publish them.

Chapter 4
Wellness as Welfare

In the forest {chimpanzees} have a great deal of freedom of choice. They can choose whether to travel on their own, in a small group, or to join large excitable gatherings. They can usually choose which individuals to associate with . . . Close companions meet often, others may avoid each other . . . To survive they must spend much time searching for and sometimes preparing their food – they are occupying their brains, using their skills. They are free.

Jane Goodall

The world's most progressive zoological parks have built their reputations with a commitment to a superior quality of life and best management standards and practices. While conservation and education are the main pillars of successful zoos, a growing emphasis on science and animal welfare is the next frontier of excellence. In Europe the connection between scientific programs and animal welfare is well-established, while animal welfare strategies are developing all over the world including recent initiatives in Australasia. Nested in animal welfare, veterinary medicine, and behavioral science, the philosophy and practice of wellness is a new concept for zoos preparing for national leadership in the domain of animal welfare. Given the strong affinity for wellness among the general population, and its familiarity to insurance companies, schools, fitness clubs, and the purveyors of health food, zoos may be the ideal venue for a comprehensive wellness initiative with appeal to families. After experimenting with the wellness idea in three American zoos, we have found it to be a highly adaptable concept that is easily communicated to all stakeholders (Fig. 4.1).

4.1 Differentiating Wellness and Welfare

The construct of wellness is understood as "a balance of mind, body, and spirit that results in an overall feeling of well-being." This definition finds expression in the constant, conscious human pursuit of living life to its fullest potential. Interestingly,

T.L. Maple and B.M. Perdue, *Zoo Animal Welfare*, Animal Welfare,

DOI 10.1007/978-3-642-35955-2_4,

Fig. 4.1 Zoo gorillas living in species-appropriate social groups that reproduce, and exhibit normal patterns of parental behavior demonstrate the principle of an optimal state of wellness (Joe Sebo)

many wellness practices are targeted to the control of obesity and inactivity, both of which are side-effects of affluence in humans, although not exclusively so. Poverty also produces obesity due to poor nutrition and inequities in health practices and opportunities. Inactive adults tend to produce inactive children and their family pets also reflect this tendency. Overweight pets suffer all the maladies that afflict people who eat too much and exercise too little. In the zoo we have to prescribe activity and diet, but we've learned diet alone is clearly not sufficient to produce animals that are healthy and well. In the psychological realm, wellness equates to fitness, and fitness is an important dimension of preventive medicine. In nature, animals keep fit to survive. In the zoo they have to be trained, and opportunities for an active life have to be scheduled daily by attentive zoo keepers. It is interesting to note that until recently psychologists have not prescribed exercise as treatment when dealing with human mental conditions even though there is plenty of evidence that exercise contributes to psychological well-being in people. In a recent review, Weir (2011) quoted Dr. Michael Otto:

> People know that exercise helps physical outcomes. There is much less awareness of mental health outcomes—and much less ability to translate this awareness into exercise action.

Mood is enhanced by exercise but there is also growing evidence that active people are less depressed. Psychologists are also investigating the connection between exercise and anxiety disorder. Just as we have to activate zoo animals, psychologists are in a position to develop new techniques to activate their patients and keep them motivated for lifestyle wellness changes. This is just one more reason why the field of psychology is a critical set of strategies and tactics in the implementation of a comprehensive zoo animal wellness and welfare program.

The lack of regular exercise may be a primary reason that so many zoo animals die from heart disease (Cousins 1979). Captive lowland gorillas develop heart disease in early adulthood, and concerned zoo veterinarians and consulting human cardiac specialists are collaborating to determine the cause and find preventive solutions to this serious problem. Although they are capable of living into their 50s, many male gorillas have died in their early twenties. The gorilla Massa died at the age of 54 at the Philadelphia Zoo in 1984. One male at the Cleveland Zoo died from heart disease at the age of 21. Alarmed by this new trend, Cleveland's scientific curator, Kristen Lukas, Chair of AZA's SSP Committee for gorillas, worked to establish a national database to study gorilla health variables in North American zoos. From this information, the role of diet, exercise, family history, and socialization will be carefully and continuously evaluated for this population. From a 1994 study of 74 lowland gorilla deaths in zoos, we learned that 41 % of the deaths were due to heart disease, primarily a condition known as "fibrosing cardiomyopathy". Seeking to intervene with appropriate medication, Cleveland Zoo veterinarians are treating two surviving males with beta blockers and ACE inhibitors, familiar treatments for human heart ailments. In Atlanta, zoo veterinarians have pioneered ultrasound examinations on gorillas while they are awake. This innovation required new technology, medical expertise from heart experts, and a training regime to render the animals compliant. Atlanta veterinarians have also worked with Georgia Tech engineers to design and develop a "Gorilla Tough Cuff" to enable awake blood pressure readings. This is a very good example of how zoos can partner with universities. Zoo Atlanta's collaboration built on an existing research relationship between the Coulter Department of Biomedical Engineering at Tech and Emory University's School of Medicine. The best science and engineering is now available to advance zoo animal wellness. Diets are also changing. At the Indianapolis Zoo, keepers do not provide fattening sweets, instead delivering sugar-free helpings of Jell-O, low-salt crackers, and green and yellow vegetables. Daily nutritional modifications are feasible now that zoos are hiring doctoral level nutritionists. Some 20 of the 200 plus accredited North American zoos and aquariums employ full-time staff nutritionists, and many others have hired nutritional consultants to monitor their diets. There is so much research in this area that the journal *Zoo Biology* is fast becoming a leader in the publication of new findings in wildlife nutrition. Our interest in zoo gorilla health and wellness will require studies of wild gorillas to learn more about their general fitness. In recent years we've discovered that lowland gorillas eat more fruit than previously believed and they have now been observed foraging for plants submerged in the swampy regions of West Africa. (Nishihara 1992; Popovich and Dierenfeld 1997; Tutin and Fernandez 1993). One of these plants, *Aframomum*, a member of the ginger family, is a powerful antibacterial, antiviral, antifungal, and anti-inflammatory food, in essence a natural medicine.

Later in this book (Chap. 6) we will provide a comprehensive review of environmental enrichment, a proven strategy for activating zoo animals. Optimal wellness cannot be achieved if animals remain sedentary. Because obese elephants experience difficult pregnancies and often require human intervention to deliver, it is important to keep them fit. Authorities in the South Indian state of Tamil Nadu revealed recently that many of the Asian elephants kept in temples are obese due to inactivity (Parameswaran, 2012). Obesity is becoming a global epidemic in captive Asian

Fig. 4.2 Burlap enrichment encourages social interaction. (T. Maple)

elephants. At the Oregon Zoo, keepers train elephants to jog for multiple 20 min sessions each and every day. The elephants also lift and push logs, and swim in their pool as directed by keepers. Both diet and exercise are elements in the regime to improve the outcome of an elephant's 22 months of pregnancy. In the wild, the need for food and water motivates elephants to move. In the zoo, keepers must endeavor to activate the elephants to keep them healthy and well. Zoos need much larger enclosures to facilitate movement in elephants, or they must creatively provide access to the entire zoo in perimeter pathways (see Chap. 8).

As we have learned, many experts regard welfare as equivalent to well-being. Mason has used the terms interchangeably (e.g. Mason and Veasey 2010). Human health expert Deepak Chopra operates a Center for Well-being where he monitors world standards, nation by nation. He uses the term "thriving" to describe optimal well-being in human populations, and "suffering" to describe its polar opposite. Because the concept of "wellness" has been deployed so effectively in medicine, especially nursing, and veterinary medicine, it may be regarded as a synonym for welfare and well-being. However, wellness may be easier to introduce as an operating standard for zoos and aquariums. So, if we can engineer the living environment and develop husbandry techniques that produce healthy, active, fit, and well zoo animals, we will achieve good or even optimal welfare. The institution that is reluctant to introduce welfare principles should be comfortable with the implications of an optimal standard of wellness. A functional wellness program will establish contingencies to activate zoo animals. With prompts from their keepers, they will learn to search and forage for food. Obesity and inactivity are correlated with the development of Type II Diabetes, a disease that was found in primates in 30 % of the institutions surveyed by Kuhar et al. (2012). We will not be surprised if zookeepers lead the way to establish wellness as an operating standard in all zoos and aquariums (Fig. 4.2).

Throughout the world, animals are still living in enclosures where activity is limited by the quantity and quality of the available space. When wellness is a priority of management, standards and practices are promulgated to ensure optimal physical and mental health for each species and every individual animal in the zoo population. The absence of stereotyped behavior and other forms of psychopathology is an important indicator of mental health. Animals that have suffered prior histories of social deprivation often can be rehabilitated with behavior modification, but prevention is always preferred over treatment. Social deficits, due to a history of isolation, are generally resistant to treatment, but there have been notable exceptions as demonstrated by the successful socialization of Zoo Atlanta's lowland gorilla, Willie B. (Maple 2001).

Proper nutrition and daily locomotion induced by spacious and stimulating surroundings, access to conspecific peers, and intervention by skilled trainers will produce good health and a desirable state of wellness (Fig. 4.3). Wellness through the science of nutrition is already a thriving business for dogs, cats and other pets. New foods are available to consumers and many are branded for wellness. For people and pets, some physicians and veterinarians are now offering treatment and counseling that reflects a spiritual, "new-age" orientation. Psychological well-being has been mediated by private practitioners who espouse their proficiency in "dog, cat, and horse whispering". On a more objective level, the comforting demeanor of compassionate keepers and veterinarians should not be underestimated as a factor contributing to wildlife wellness at the zoo. Recent research on human happiness may be useful in our quest to understand wellness in animals. Writing in the journal *Perspectives on Psychological Science*, Oishi et al. (2007), used the term "happiness" interchangeably with the construct "subjective well-being." They regard happiness as a relatively stable feeling, rather than a momentary, transient emotional state. From their research on human subjects, the authors concluded that extremely high levels of happiness might not be a desirable outcome. Indeed, a moderate level of happiness may be regarded as optimal, since extreme happiness might lead to dangerous thrill seeking or instability in social or work relationships. This idea can be traced to Aristotle who proposed that people should cultivate virtues at mean or intermediate levels between deficiencies and excesses (Grant and Schwartz 2011). One can extrapolate here to argue that a zoo animal might be optimally "happy" or well even if it could not hunt, fight, or flee from danger in a zoo habitat. This may also be an argument for the application of the Yerkes-Dodson Law (the empirical relationship between arousal and performance), so useful in describing the value of steady emotional states during competition and conflict. In this context, high levels of arousal should be avoided.

4.2 Practicing Wellness

Wellness is based on the standard practices of preventive medicine as zoo professionals work cooperatively to keep animals healthy and free of disease and injury. Medical personnel use training techniques to reduce the stress of intervention. Wellness is therefore a philosophy inherent to the partnership between zoo medicine and animal care and husbandry. Zoo keepers, trainers, and medical technicians operate

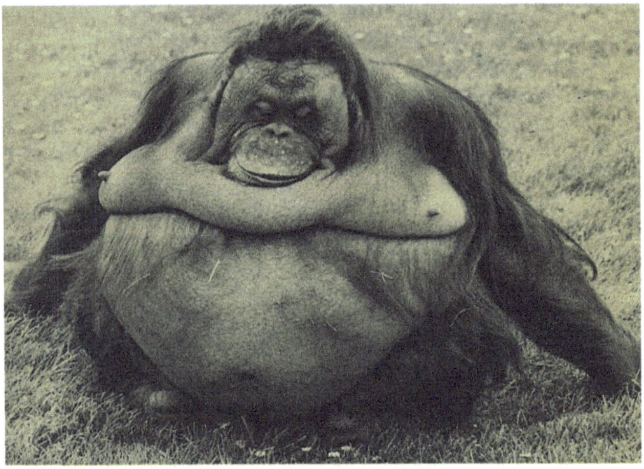

Fig. 4.3 Inactive zoo orangutans in the zoo are prone to suffer obesity and Type II Diabetes

on the frontlines of wellness. In a zoo, achieving wellness can be very labor-intensive and must be carefully planned. Advocates of wellness and welfare are generally comfortable with the dichotomy of poor v. good welfare. However, if nature is the model for wellness and welfare, it is possible if not preferable to aim a little higher and try to achieve or approach a state of "optimal wellness". The standard of "high level wellness" has been used by early wellness practitioners in human medicine. Distinguished from the term "health," the term "wellness" shifts attention from illness to a state of well-being and its ongoing development. The Singapore-based National Wellness Association defined it for humans as "an active process of becoming aware of and making choices toward a more successful existence." The Herbert Wellness Center at the University of Miami recognizes seven types of wellness that are germane to formation of a "balanced" life style for people: (1) Physical; (2) Intellectual; (3) Emotional; (4) Social; (5) Spiritual; (6) Occupational; (7) Environmental.

To reiterate, with captive animals, the opportunity to be healthy has to be managed by human intervention and creative environmental engineering, although there are opportunities to provide choices for captive animals as Markowitz and others have repeatedly demonstrated. In the zoo, keepers, curators, and veterinarians are ultimately responsible for the animal's standard of wellness. With each species, a behavioral ethogram can be developed to define the applicable scope of wellness. If we understand the key variables that influence locomotion, attention, consumption, socialization, play, cognition, and reproduction, we can evaluate and monitor species-typical behavior and hopefully achieve optimal wellness. This is another reason why a dedicated doctoral-level animal behaviorist is a valuable colleague on the animal care team. Wellness begins with a deep understanding of the preferences and the needs of each and every species in the zoo. Therefore, zoos committed to wellness and welfare must recruit intelligent, perceptive employees with advanced training in relevant fields such as psychology and biology. To a person, they must know or come to know the animals entrusted to their care, and protocols must be developed to fit the unique, species-specific attributes of the animal.

4.3 Managing Mega-fauna

Wellness is especially challenging when attending to the needs of the largest creatures in the zoo. At the Palm Beach Zoo in Florida, two fully adult alligators (*Alligator mississippiensis*) developed sedentary habits to an extreme. At 13 ft in length, the male had reached the weight of 900 lb, or more than 10 % greater weight than a typical, healthy alligator. As part of an overall wellness husbandry program, both animals were selected for special training and a high protein diet. Using small pellets of food to reward compliance, the alligators were trained to move about the pool system by verbal prompts and food rewards dropped into their pool by zoo keepers. The animals quickly learned to exercise in the enclosure for food rewards, and with the reduced diet soon began to show greater activity associated with hunger. In fact, keepers were surprised when the male alligator suddenly turned on a bird that had occupied its enclosure without incident for many years. The bird (a night heron)) was attacked and eaten in front of keepers and a surprised gathering of visitors. This led to the formulation of a local corollary to wellness; "wellness has consequences!" With time, the male alligator visibly lost girth although he was not weighed to confirm the loss. He remained more active on exhibit, began to hunt and fish without prompting, and learned to exit the pool to walk on land by voice command. Keepers were able to demonstrate the learning skills of alligators and discuss the wellness paradigm to the delight of fascinated visitors. In this case, alligators proved to be an ideal subject for teaching about wellness (Fig. 4.4).

At Disney's Animal Kingdom in Orlando, Florida, veterinarians periodically conduct a wellness examination on their Nile crocodiles (*Crocodylus niloticus*). The park currently manages 26 adult male crocodiles. When the exam is on, the animals are lined up in a long series of enclosed chutes in a riverine system behind the scenes. One by one the animals enter the exam chute where they are secured so staff can draw blood and carry out other medical tasks. Once each animal has been examined, the front door is opened and a processed whole chicken is launched, followed by a leaping, ravenous crocodile now liberated to consume it. This is the most fascinating demonstration of wellness medicine for any animal that we have witnessed in a zoo. Although Disney carries out this procedure behind-the-scenes, it is a splendid example of how wellness could be taught by sharing zoo animal medicine with the public. A wellness examination may be considered an enrichment experience, although they tend to be infrequent for some species reducing its value for public education. As the animals would likely work to participate, perhaps a modified procedure could be carried out by keepers or the entire ordeal could be filmed for used in a wellness education center. In such ways, respect for wildlife is engendered. After witnessing a training session at the zoo, we often hear visitors exclaim, "gators are smarter than I thought". Respecting alligators is important in

Fig. 4.4 Alligator at the Palm Beach Zoo awaiting positive reinforcement (E. Maple)

Florida since they are abundant and potentially dangerous to humans who come too close.

4.4 Integrating Wellness

Every department in the zoo has a stake in the organization's wellness and welfare strategy. An organizational structure that supports wellness intervention should carefully evaluate human and financial resources to determine how optimal wellness can be achieved. The noble idea of wellness will create opportunities for the advancement team to raise money from new sources, locally and nationally. Wellness is an especially attractive opportunity for foundations that don't normally give to zoos. Educators can be deployed to teach about wildlife wellness in the schools both to illustrate superior zoo standards and practices and to obliquely teach children that good health practices are important for people and animals alike. Animal stories are subtle prompts for kids who do not suffer gladly the criticism of adults. We have learned that children are fascinated by the notion that sedentary male gorillas are prone to the early onset of heart disease, that obese orangutans develop type-II Diabetes, and that overweight female elephants may require human medical intervention to deliver their babies. A recent media example of the power of animals to educate children featured a morbidly obese dachshund in Portland, Oregon, undergoing therapy to lose weight. The owner expressed her hope that people suffering obesity would be inspired by the dog's story. In her doctoral dissertation, Stephanie Allard found that school teachers in Palm Beach County, Florida understood and appreciated the promise of the wildlife wellness model and the potential of zoo stories to deliver a compelling message to children. A wellness partnership between zoos and school systems might be marketable as an integrated

Fig. 4.5 Crocodile wellness exam at Disney's Animal Kingdom (Courtesy Walt Disney Company)

national program with the help of a significant sponsor. A simple way to introduce the concept to children is to modify the traditional zoo-mobile and send it on the road as the zoo "well-mobile" with small animals playing the role of wellness ambassadors. The wellness brand can fit in virtually all educational settings (Fig. 4.5).

Anyone who deals with the visitor, from volunteer docents to keepers and curators, will be needed to talk about the zoo's comprehensive commitment to wellness and point out the many features of the construct that are unique to each species. Public Relations and marketing staff will have the opportunity to build a case for wellness on the website, in periodicals, and for other forms of social media carrying the message to zoo members and friends. In-park signs will carry new messages that explain wellness and how it is shaping the life of each animal and each new exhibit. The story of wellness will be revealed in case studies that will generate excitement and enlarge the circle of friends and donors for the zoo. Detroit Zoo employees and volunteers have mastered the art of promoting their animal welfare program to their guests, their community and visiting dignitaries. This should be a model for any institution that seeks to raise the profile or wellness or welfare.

4.5 Branding Wellness

It will be the task of marketing and public relations staff, zoo leadership, and consultants to find the right vehicles and venues to locally and nationally market the new wellness brand. To be "First in Wellness" among the nation's best zoos, the

Fig. 4.6 Hedgehogs (*Erinaceinae*) are compelling ambassadors for wellness (K. Cytacki)

branding plan must be sophisticated, ubiquitous, and persistent. Any time an organization develops a strong brand that elevates its reputation, it attracts new partners and sponsors. The zoo should look for credible local and national corporations that have already invested in wellness (e.g. insurance companies, hospitals, grocers). Local, regional, or national health food stores and organic food producers are good prospects to promote wellness in the zoo or in joint public marketing campaigns. The corporate or professional partner can provide funding to widely promote wellness in the zoo or in joint public marketing campaigns. Wellness is a powerful theme that has the potential to link every component of a zoo's vision and philosophy.

As an institution plans for the elevation of wellness to a position of priority, zoo staff should examine the zoo population to plan for future changes influenced by strategic master-planning. A strategic animal population (collection) plan should examine each grouping of wildlife, species by species, group by group, exhibit by exhibit, to identify the best opportunities to succeed. Ideally, a plan for successfully exhibiting and managing each species can be achieved in the context of both new exhibits and exhibit renewal and renovation. Strategic planning will identify design and development opportunities from many perspectives including the new dimension of wellness. Any existing strategic plans should be adjusted to acknowledge wellness as a business strategy and a key priority in the zoo's mission (Fig. 4.6).

Zoo staff and outside consultants should collaborate on the application of sensitive metrics to evaluate animals and identify vulnerabilities (Chap. 3). From this process, priority species can be identified for medical or behavioral intervention. Equally useful are surveys of visitors, staff, volunteers, and members, to determine priorities based on public perceptions of wellness. We have provided some sample questions from a wellness survey in Table 4.1. Wellness and welfare metrics are essentially the same. Animals that are not living well should be elevated to the highest priority for change. Zoo veterinarians routinely conduct wellness exams to determine an animal's total condition. Just as people visit their doctor for an annual physical, the annual

Table 4.1 Sample Wellness Questionnaire (Courtesy S. Allard)

1. Check all items that indicate wellness in zoo animals:
• Animal is active.
• Animal appears fit.
• Animal walks with a normal gait.
• Animal is alert.
• Animal interacts with others.
2. Check all items that indicate a deficiency in wellness.
• Animal is behaving abnormally.
• Animal is pacing back and forth.
• Animal appears to be afraid.
• Animal has too little space.

wellness exam checks the weight and all measurable physical Parameters that indicate good or poor health. To conduct wellness exams, zoos must be engineered to encourage evaluation. Enclosures must be fitted with appropriate scales, squeeze devices, or capture chutes, and animals, as exemplified by the Disney crocodiles, can be easily trained to accept non-invasive medical procedures. Using operant conditioning techniques, many animals have learned to cooperate for blood pressure readings, and veni-puncture without anesthesia. The psychological dimensions of wellness have proved to be more difficult to evaluate, but the key to understanding the psychology of welfare is systematic observation.

In 1996, demonstrating the efficacy of using training techniques to practice veterinary medicine, staff at the Whipsnade Zoo utilized operant conditioning techniques to shape greater one horned Asian rhinos to accept veterinary foot-care. Two years later they trained black rhinos to tolerate blood collection and ultrasound examinations without anesthesia. Rhinos proved to be compliant and cooperative in adapting to medical intervention. As we will explain in great detail in Chap. 6, optimal wellness (and welfare) is greatly facilitated by compliance training and the concomitant reduction of stress when heavy-handed capture techniques are no longer practiced. Any attempt to reduce anxiety and fear is a contribution to improved wellness and welfare (Fig. 4.7).

4.6 Dedicated Wellness Centers

In simpler times zoos built "commissaries" to store, prepare, and distribute food for the animals. The commissary disappeared from the zoo vocabulary when nutrition emerged as an important specialization in the zoo. Many of the bigger American zoos, e.g. Bronx, Minnesota, National Zoo, San Diego, St. Louis and others gave a boost to this trend by recruiting doctoral level colleagues in nutrition. A good example of the upgrade is the Orthwein Animal Nutrition Center at the St. Louis Zoo. In addition to food storage and preparation, this facility includes a fully equipped laboratory and work space for serious nutritional studies. The building has been equipped for advanced nutritional research but it is also accessible with large windows that encourage zoo visitors to observe the workers inside. Visitors

Fig. 4.7 Adult Sumatran tiger trained to stretch for full body inspection at Zoo Atlanta (A. Thompson)

are fascinated by the volume of food consumed by zoo animals and they like to help when they can. One of the most popular volunteer jobs in the zoo is daily diet preparation. Although diet and nutrition in the zoo is a scientific discipline, the psychological aspect of food preparation and food service is equally important.

Advanced nutrition units are often combined with veterinary hospitals and take the form of a comprehensive wellness center as both the locus of nutrition research and the traditional tasks of food preparation and delivery. A wellness center should also be the location where enrichment material is prepared and distributed by keepers, educators and zoo docents. In this way, a designated wellness center confronts the full spectrum of health and wellness, both biological and psychological elements. By locating enrichment preparation in the wellness center, guests are able to experience the excitement of distributing prized browse to a diversity of zoo animals. A full-service wellness center can be managed by the Chief Veterinary Officer or by other senior animal care administrators.

The Virginia Zoo, located in Norfolk, recently announced a campaign to build a "wellness campus" comprised of an 8,000 sq. ft. veterinary hospital and diet kitchen and a venue for educating the public about animal care and medicine, and the benefits of good nutrition, outdoor activity, and exercise for visitors. The campus will also attempt to teach how local food is grown and harvested, and how food for zoo animals is prepared. Norfolk's wellness campus will be a creative example of a comprehensive wellness program, and an opportunity to promote wellness ideas at the zoo. A wellness center has a slightly different purpose from one dedicated to welfare, with its origins in health and medicine, and its natural affinity to veterinary medicine, but welfare and wellness are synergistic. The idea of a wellness campus has design implications as it is another way the zoo can contribute to human health and wellness.

Environmental psychologists have recently focused their attention on promoting exercise in public venues (Brown et al. 2007). A zoo can be set up to provide long strolls among the botanicals and the naturalistic animal enclosures. A walking safari through a "pedestrian-friendly" landscape immersion zoo is a meaningful contribution to the visitor's health and wellness. This is comparable to the recent trend in designing buildings that activate people as discussed in Chap. 8. Where buildings used to provide fast access to the far reaching levels of the structure, healthier designs encourage people to walk to their destination with exposed stairwells and other features.

Designing for welfare/wellness is primarily an exercise to improve the living conditions for captive animals, but the opportunity to educate visitors and even to improve their own sense of well-being or wellness is a good reason to feature interpretive kiosks and other forms of messaging throughout the zoo. In addition, a dedicated wellness center gives management an opportunity to demonstrate how enrichment programs are carried out on a daily basis by zoo keepers. A story being told at the Denver Zoo is instructive. Visitors often ask employees whether the animals know their keepers as individuals. The answer, of course, is absolutely, they do know their keepers. When Craig Piper, the zoo's CEO, was asked to comment, his succinct answer spoke volumes:

> Every relationship between keeper and animal is unique. That bond is so important. It's the trust, the cooperative spirit that enables us to take care of them.

Keepers are so essential to the achievement of optimal wellness that their personal relationship to the animals in their care may be the best zoo story of all. The unique keeper-animal relationship is also a way to showcase the impressive cognitive skills of great apes and other mentally advanced species. A long history of research and husbandry experience has demonstrated that cognitive enrichment is a powerful management practice. Video technology can be installed to provide for two-way communication between animals and their keepers. We wonder how great apes would respond to the opportunity to view familiar keepers preparing food, as the ability of primates to learn from video-tapes is well –documented (Washburn and Rumbaugh 1992; Bloomsmith and Lambeth 2000; Rumbaugh and Washburn 2003). Visitors to a well-equipped wellness center would have the opportunity to keep an eye on both venues, while zoo biologists could study the reactions of apes and visitors. We suspect that the great cats (lions, tigers, leopards, and jaguars), would also respond to video-taped images of food preparation. This would be another interesting assignment for the Imagineers at the Walt Disney Company (Fig. 4.8).

A wellness center should also be the venue where future exhibits are described and illustrated with colorful, artistic and architectural renderings. The story of how animal welfare is influencing future exhibits is worth telling and it brings the visitor into the discussion. At this location, the visitors could be asked to provide ideas for future projects; how would they improve welfare and wellness for a diversity of animals in the future zoo? Every zoo should have a zoo version of a "real estate" sales center where visitors can carefully learn about the plans for exhibiting different species and different zoo-geographic regions down the line. Here they can voice their support, express their preferences and opinions, and ask questions about the priority facilities on display. The Wellness Center is an appropriate place to recruit new members and donors and, sometimes, new ideas.

Fig. 4.8 Natural behavior contributes to wellness, offering teachable moments (Detroit Zoo)

Normally, nutrition and veterinary facilities are placed at the perimeter of the zoo campus, but a wellness center is best utilized if it can be sited at the center or even at the entrance to the zoo. The idea is to share the priority of wellness and welfare with zoo visitors, and to utilize wellness as a strategy to influence the health and welfare of families and communities. The zoo's wellness center should be the dynamic centerpiece of the zoo's commitment to wildlife welfare, and in this way its public service function is dramatically expanded. By locating a wellness center at the entry-exit point of the zoo, the psychological principles of "primacy" and "recency" are deployed as strategies of design. As visitors enter and depart, the wellness center becomes the first and the last impression of the visitor's zoo experience and therefore more likely to be remembered. What other influence might the center have on the behavior of visitors? It would be very interesting to learn if the power of the wellness educational experience would influence the visitor's choice of food. If its location was contiguous to a restaurant or café, the mix of healthy versus traditional zoo food might be dictated by the lingering effects of the wellness message. By telling our visitors how we improve zoo animal health through proper nutrition, we might see immediate results in the food choices of our visitors when they purchase their lunch at the zoo.

In human health circles, the wellness center is becoming ubiquitous. On most college campuses, wellness centers have replaced infirmaries, while pet owners can now take their dogs and cats to veterinary wellness centers. The University of California at Davis (UCD) College of Veterinary Medicine and many others throughout the world are developing wellness research programs and deepening their expertise in animal welfare for animals in agriculture, biomedicine, and zoos. An exotic parrot wellness specialty at UCD was developed to recognize the special needs of these long-lived and complex family companions. Gyms and exercise facilities have also begun to rebrand their industry in terms of "health and wellness". This trend is advancing so rapidly it is surprising that more zoos and aquariums have not adopted wellness terminology as an operating philosophy, but we expect the wellness concept to be adopted widely as a strategy for advancing zoo animal welfare.

Although wellness programs have been adopted for captive primates in bio-medical facilities (e.g. Fritz et al. 1993), wellness is a relatively new management concept for zoos and aquariums. As we have seen, zoo veterinarians are trained to conduct wellness physical exams on a variety of animals, and domestic animal veterinarians have embraced the wellness concept with enthusiasm. Private veterinary clinics boldly market their commitment to animal health and wellness, but zoos have been more cautious. It is the behavioral side that has not been fully integrated with health and wellness management in the zoo. Some zoos have invested in behavioral management in a big way while others have outsourced the work to consultants such as "Active Environments" (e.g. Laule and Desmond 1998). Wellness is the paradigm that will bring these two approaches, the biological and the psychological, into correct alignment. This is not a false dichotomy; they are in fact specialties that deserve equal emphasis and demand differentiated expertise. Although veterinarians commonly acquire additional training in animal behavior, their work in wellness is enhanced by partnerships with ethologists and psychologists.

4.7 The San Francisco Wellness Initiative

In the fall of 2011, leaders at the San Francisco Zoo began to explore a formal and comprehensive "wellness initiative" to discover new techniques for improving the living conditions of each and every animal in the zoo. In a concept paper developed in 2012 by zoo management, the writers proclaimed:

> A zoo worthy of the name celebrates all forms of life living to their full potential. Such a zoo promotes physical and psychological well-being for animals and humans, as well as health for the environment on which all life depends. The term 'wellness' embodies these broad aims. For this reason, the San Francisco Zoo inaugurates the Stanton Family Wellness Initiative to emphasize our commitment to life and well-being for our animals, our community, and the environment.

The wellness concept conforms to the prevailing San Francisco Bay area intellectual climate where a culture of fitness and wellness, and a plethora of interest groups and businesses promoting organic foods, nutritional supplements and alternative medicines, and spiritual systems for restoration and renewal are ubiquitous. Many of the zoo's visitors expect to be able to purchase a vegetarian meal from local sources, and they want to see healthy, active, and contented animals in the zoo. The former expectation was facilitated by the "Leaping Lemur Café" opened in 2002 with plenty of fresh, local, organic and vegetarian food items to choose from, and a decidedly different type of zoo restaurant. In San Francisco the new restaurant has been wildly successful, although patrons still consume a lot of hot dogs and hamburgers. A full-service approach to wellness is inspiring on many levels.

Wellness can be achieved through the application of immediate management techniques and training, and implemented long-term by designing wellness-oriented facilities where animals are activated and stimulated by species-appropriate environmental and social conditions. Currently there are several exhibits at the San Francisco Zoo that are already built for optimal wellness including a highly acclaimed lemur forest dispersed over 1.4 acres of naturalistic terrain. Five lemur species are featured. The Lipman Family Lemur Forest is comprised of Cypress and Eucalyptus trees reaching 70 ft in height. Sixteen of the twenty lemurs exhibited are males which encourages competitive leaping and climbing and a continuous cacophony of vocal displays throughout the day. The lemur habitat is both environmentally and socially complex. There are two artificial climbing structures with heated platforms, basking lamps, and four heated nest boxes. An elevated viewing boardwalk for guests enhances the immersion effect of the exhibit. On surveys, visitors have consistently given the lemur forest high marks for its welfare features.

Another exhibit under consideration in San Francisco provides a spacious vertical habitat for a male group of squirrel monkeys, animals that were rescued from a local biomedical lab. This facility provides an adjacent play area for children where they can mimic the arboreal antics of the monkeys, another good example of parallel wellness. There are many other restorations under consideration in San Francisco including a gorilla enclosure that was once one of the nation's best. After 30 years of service, the wellness initiative provides a reason for revitalizing this facility (Fig. 5.7). As the exhibit was designed with 360 ° perimeter viewing and a topography that locates visitors above the gorillas, looking down at them, the effect is suboptimal. Both features are design flaws according to Jones (1982) and Coe (1985; Chap. 8) Viewing areas have been modified to discourage barrier violations by visitors and some lookouts have been planted to reduce the totality of intrusions at the perimeter. These modifications are examples of wellness-inspired design. Many of the holding buildings at the San Francisco Zoo are classic examples of hard architecture, but these structures can be improved by short-term renovation or a complete re-design. In addition to gorillas and chimpanzees, top priorities for wellness-inspired design are the zoo's great cats (lions, tigers, leopards) and polar bears. The prospect of achieving enhanced wellness for beloved and charismatic mega-vertebrates is a very exciting development (Fig. 4.9).

Fig. 4.9 Naturalistic lemur habitat at the San Francisco Zoo (M. Woon)

Happily, the zoo's grizzly bears (*Ursus arctos horribilis*) are already living in a one-acre enriched environment, Hearst Grizzly Gulch, opened in 2007. In their new $3.7 M habitat the two bears can feed on live fish in a 20,000 gal shallow pool, and enjoy a lush meadow, heated rocks, a two-ton tree stump, a pit for digging, and an herbal garden under cultivation for bears only. The two bears in this exhibit behave like wild bears, standing on their back feet to visually search for fish, diving into the water to capture them, and consuming as many as they can hunt in a feeding frenzy that is wonderfully educational. Now that they have witnessed optimal habitats and their enriching effect on grizzly bears, members, visitors, and zoo staff welcome the opportunity to provide a state-of-the-art polar bear facility in San Francisco. According to Clubb and Mason (2003) polar bears need wide open spaces to roam. Because they live in zoo exhibits that are a million times smaller than their natural home range, they are prone to develop serious pacing stereotypies. Polar bears are good swimmers and like to dive in deep pools providing visitors an opportunity to observe their athleticism in water, and they must be challenged with a changing environment to compensate for the loss of opportunities for locomotion. They can be provisioned with both live fish, and fresh fruit, and vegetables, which rewards them for regular exercise and exploration. So far, there have been no objections to the periodical feeding of live fish to grizzly bears. In a Swiss study germane to this issue, zoo visitor attitudes toward live feeding were surveyed. Cottle et al. (2010) confirmed that the use of species deemed higher on the phylogenetic scale generate concern. Feeding live rabbits to zoo tigers, for example, runs the risk of compromising the visitor experience. Fish seem to be universally accepted as live food for zoo animals and clearly contribute to the psychological well-being of bears. However, a recent book by Victoria Braithwaite (2010), a Penn State University biologist, has concluded that fish do indeed feel pain and suggests that new information about fish requires better treatment by individual anglers and commercial fishing industries. Grizzlys evolved to pursue, capture, and consume fish in nature, but the use of live fish as enrichment in the zoo will likely be debated in the near future. Because bears represent a challenge to the

Fig. 4.10 Grizzly bears playing in water at the San Francisco Zoo (M. Ransone)

zoo designer who must create highly stimulating surroundings in a complex and natural space, we surmise that any welfare calculus is likely to conclude that bears have priority over fish (Fig. 4.10).

4.8 Teaching and Outreaching

In a dedicated wellness center and throughout the zoo, in educational kiosks and through creative and colorful signage, the zoo's compelling stories of wildlife wellness can be shared with the public. By linking human and wildlife wellness through stories, the zoo becomes a relevant contributor to the health and well-being of our communities, and a stakeholder in environmental wellness. Many years ago, the San Francisco Zoo was known for its popular lion and tiger feeding sessions in the cavernous feline building that is still standing. Here keepers would deliver hunks of meat to the lions, tigers, and leopards to the public's delight. Roaring was the order of the day so the full power of the great cats was on display. An anachronism that no longer exists in many zoos, the wellness exam, creatively construed, is the modern day equivalent of the great cat feeding experience. Zoo keepers, working with well-trained felines, can easily get them to show their great size to the visitor with just a click or a verbal prompt. They will also open their mouth to expose their powerful teeth, a dental exam that is most impressive. Vocalizing can also be taught and here, too, the animals are provided with food rewards so it still resembles the old-school feeding sessions, but more controlled, and carried out for a nobler purpose, the advancement of the animal's health and psychological well-being. The new wellness examinations are now a regular feature of the zoo keeper's daily interactions with the great cats at the San Francisco Zoo. It is clearly an experience that is beneficial to the animals, zoo keepers, medical staff, and zoo visitors. The building may be old, but the San Francisco Zoo's action-oriented wellness program is entirely new.

Other zoos practice this form of interaction in keeper demonstrations with a variety of animals (e.g. Fig. 7.5) and the result is both educational and entertaining.

Wellness will be perceived as equivalent to welfare by our supporters and our critics, but wellness doesn't carry the political baggage of welfare, and may be a quicker, less contentious path to promoting and implementing animal welfare in zoos and aquariums. Both wellness and welfare are core values that support enhancements to the living standards of zoo and aquarium animals. Wellness is a condition we share with animals, and our growing awareness of what it takes to be well is helping us to extend this outcome to every animal at the zoo.

Chapter 5
Psychology and Animal Welfare

> *It is hardly an exaggeration to say that a chimpanzee kept in solitude is not a real chimpanzee at all. That certain special characteristic qualities of this species of animal only appear when they are in a group, is simply because the behavior of his comrades constitutes for each individual the only adequate incentive for bringing about a great variety of essential forms of behavior.*
>
> Wolfgang Kohler

By now the reader is aware of the core values and professional biases of the authors of this book. We are academic psychologists with fundamental training in animal behavior, behavior analysis, comparative cognition, and environmental psychology. Although we have collaborated to study a diversity of mammalian species, our primary expertise is derived from studies of nonhuman primates, especially the great apes. Our scientific network operates in both the academic and the zoo world. After decades of basic and applied research, we are convinced that zoo animal welfare cannot succeed without active collaboration with individual scientists and universities. In the senior author's capacity as a non-profit executive, we have also had the responsibility of negotiating partnerships and affirming research relationships. We believe that zoos with scientific credibility can be trusted to do the right thing on behalf of the animals. Our confidence is sustained by our experience that managing animals is comprised of both scientific and intuitive elements that sometimes compete for dominance in the zoo. To keep management on the empirical side, evidence-based managers, scientific staff, and outside collaborators should be at the table when important decisions are made.

5.1 Psychologists in the Zoo

Although the prolific Swiss zoo director Heini Hediger studied animal psychology, it is an unusual background for a zoo director. In his iconic book *Man and Animal in the Zoo*, Hediger revealed the extent that the field of psychology had influenced his

T.L. Maple and B.M. Perdue, *Zoo Animal Welfare*, Animal Welfare,
DOI 10.1007/978-3-642-35955-2_5, © Springer-Verlag Berlin Heidelberg 2013

thinking about wildlife in the zoo. Like Hediger, we have found graduate-level training in psychology to be extremely useful in all aspects of zoo work. Because a zoo is not just for animals, the traditional scope of psychology provides perspective on both animal and human behavior. Indeed, Hediger recognized that the principle challenge for the captive animal is its relationship to humankind, and he made this a specialized topic within the interdisciplinary field of zoo biology. It can also be said that wildlife conservation has as much to do with people as it does with animals. While poachers and exploiters put animals at risk, dedicated conservationists work with local people to intervene and protect wildlife. An understanding of human attitudes and human values will help us solve conservation and welfare problems. An entirely new generation of ethologists and comparative psychologists throughout the world has recently entered the zoo profession as scientists, curators, educators, and administrators. These are the stakeholders who will lead the advance to an enlightened animal welfare science and practice.

Psychological science can be defined as the "study of behavior and mental life". Until recently the field was dominated by the school known as "behaviorism" which provided the intellectual framework supporting environmental enrichment, training, and behavioral management and husbandry. Cognitive psychology has replaced behaviorism as the leading edge of psychological research with animals and humans, while cognitive neuroscience has opened up new strategies for investigating mental function. Now that we can train any species to cooperate and comply without anesthesia or invasive, painful procedures, we can ask questions that were impossible just a few years ago. Oddly, with greater access to zoo collections than ever before, and with new high powered technology, animal research has nearly disappeared from the psychology curriculum at major universities. The promising fields of comparative cognition and cognitive ethology will need a boost from university administrators in order to take advantage of a diverse population of subjects residing in zoos and aquariums, and to take advantage of the growing public interest in wildlife. The cognitive capacity of many species has not been studied by psychologists, so there is a world of opportunity for current and future zoo and university collaborations. Psychological science, broadly applied, may have more potential to advance zoo standards and practices than any other scientific discipline. To tap this potential we recommend an organized recruitment of psychological scientists, including cognitive, comparative, developmental, and social psychologists. All of these specialties have something to contribute to zoo animal welfare. It is fortunate indeed that psychologists are greatly valued by our colleagues in colleges of veterinary medicine. Collaboration between these two disciplines is likely to yield impressive results and should be encouraged. The zoo is the venue where many of these successful collaborations have already produced results (Fig. 5.1).

In the long history of comparative psychology the welfare of animals was an early topic of interest since it was difficult to keep animals healthy and alive in the days when little was known about their biology or their needs. Psychobiologist Robert M. Yerkes studied great apes in captive settings long before he succeeded in starting the first primate center affiliated with Yale University and located in Orange Park, Florida (Maple 1979b). Yerkes studied the gorilla Congo in the Ringling Brothers

Fig. 5.1 Orangutan selects familiar face in cognitive study at Zoo Atlanta (A. Clay)

Circus (Yerkes 1927), the orangutan Julius at the invitation of G.V. Hamilton (Yerkes 1916), a colleague living on the grounds of a private residence in Montecito, California, and chimpanzees at the Cuban estate of Madame Rosalia Abreu (Yerkes 1925). Yerkes' iconic book *Chimpanzees: A Laboratory Colony* (1943) is filled with valuable information about chimpanzee husbandry, management, and behavior. Many scholars acknowledge Yerkes for his early interest in the subject of enrichment (Adams 2007). His words have a contemporary ring and reveal his holistic perspective on primate well-being:

> The physical environment should be designed to afford also adequate facilities for freedom of motion, exercise, play and interesting occupations. All this is merely the necessary substitute for what nature demands of the animal, if it is to achieve a livelihood and survive in the African wilds. To meet this general need space alone is insufficient, for there must be things to play and work with, perceived or compulsory ways of working to get food, companionship, amusement, or other incentives to effort. To keep an undomesticated primate, and especially a chimpanzee, captive without intelligent attempts to provide occupational filling for its idle hours as insurance against ennui or depression is entirely inexcusable. (pp. 200–201)

Psychologists associated with the Florida field station of the Primate Laboratory of the Yale Institute of Psychobiology were the first scientists to study great apes in the field (Maple et al. 1979b). Yerkes sent his associate C. Ray Carpenter to Asia to study the orangutan, while his colleagues Henry W. Nissen (1931) and Harold. C. Bingham (1932) studied chimpanzees and mountain gorillas in Africa respectively. These pioneering field studies took place long before any anthropologists studied monkeys and apes in the wild. Carpenter, a social psychologist by training and a Stanford classmate of Harry F. Harlow, also wrote a benchmark paper on the behavior of two rare Eastern lowland gorillas (Mbongo and Ngagi) residing at the San Diego Zoo (Carpenter 1937). Carpenter's innovative study stimulated research at the Bronx Zoo on lowland gorillas (Riess et al. 1949). These writers extolled the value of zoos as training facilities for field workers:

> Within the reach of researchers in most large cities there exist collections of living animals in great variety of species and under varied living conditions. The reference is to the zoological parks and exhibition areas. In many of these, natural habitat conditions are approximated and even the differences can be fertile sources of comparative psycho-ecological studies. (p. 111)

Findings from the published literature on animal behavior, past and present, ought to be the foundation for best practices in animal management and animal welfare. Needless to say, research from both zoo and field biology is equally valuable in the design of welfare protocols. To develop superior programs and superior facilities, a zoo or aquarium needs to talk first with experts who know the animals and the ecosystems that are represented in the exhibit. While it is easy to find generalists who can develop guidelines, the first step should be a visit with experts on the front lines of ecology and conservation. The world's most trusted field scientists are usually accessible to zoo directors and zoo designers. Distinguished biologists such as E.O. Wilson, George Schaller, Alan Rabinowitz, Jane Goodall, and Frans de Waal, to name just a few, have either worked in zoos or demonstrated interest in them. Aquariums have traditionally relied upon experts in oceanography to understand the biology of specimens captured for exhibition. Ideas generated at this level of expertise can drive a vision that will capture the interests of the public and lead to innovations in daily management. We continue to believe that thinking big (that is "great") is the key to achieving optimal zoo animal welfare. Occasionally, field scientists venture into the zoo literature with a contribution that explains how field data informs management and design for the animals they know best (e.g. Chap. 7). Some of the world's most impressive wildlife exhibits have not been costly, so thinking big about ideas and concepts can be accomplished in a cost-effective way.

A fundamental difference in the positions taken by animal rights vs. animal welfare advocates is their willingness and propensity to utilize sound scientific findings. Animal rights groups routinely resort to hyperbole, while credible animal welfare organizations are more likely to support their positions with evidence. Therefore, it is not surprising that respected animal welfare organizations have formed useful partnerships with scientific and medical personnel from institutes and universities to establish data-based, peer-reviewed publications such as the *Journal of Applied Animal Welfare Science*, and *Animal Welfare*. UFAW, the Universities Federation of Animal Welfare in the United Kingdom is a good example of how academics have partnered with government, industry, and non-profits to the benefit of captive animals. UFAW supports science and technology that advances animal welfare, organizes conferences, publishes media and reports, and a peer-reviewed scientific journal. UFAW is a consistent and credible source for findings pertinent to zoo animal welfare. The Scientists Center for Animal Welfare (SCAW) is another valuable website for ideas, news, and information (www.scaw.com).

In addition, traditional scientific journals with a broader audience are beginning to accept contributed papers with data that bear on bioethics and animal welfare. The journal *Zoo Biology* from its inception in 1982 has encouraged empirical studies of environmental enrichment and animal welfare (e.g. Glatson et al. 1984; Lindburg 1988; Shepherdson et al. 1993; Swaisgood and Shepherdson 2005). Under the editorship of Gordon Burghardt, the *Journal of Comparative Psychology*, a publication of the venerable American Psychological Association, recently began to solicit manuscripts on the "behavioral biology of conservation and animal welfare" (American Psychological Association Online; *Journal of Comparative Psychology*; journal description). *Nature* has also published animal welfare research documenting the vulnerability of certain species of wide-ranging carnivores that apparently experience greater suffering in captivity (Clubb and Mason 2003). An excellent example of empirical animal welfare is the recent meta-analysis of enrichment data for 54 studies published in a variety of journals (Shyne 2006). Her data analysis published in *Zoo Biology* provided powerful confirmation that enrichment can reduce the frequency of stereotyped behavior exhibited by mammals living in zoos. The journal *Science*, widely circulated around the world by the American Association for the Advancement of Science (AAAS), recently published a paper by Ross et al. (2008) who argued that media portrayals of chimpanzees rendered them more appealing as pets thereby compromising their conservation status. Both *Science* and *Nature* have assigned reporters to cover contentious public debates about animal welfare and animal rights.

5.2 Ethological Standards for Animal Welfare

The historical association of animal behavior and animal welfare was reviewed by Harold W. Gonyou (1994). In the fields of agriculture and veterinary medicine behavior research goes by the label "applied ethology". Gonyou noted that a classic book on the behavior of domesticated animals never used the word "welfare" in the entire 600 pages of two editions (Hafez 1962; second edition, 1969). Instead, the terms "mental and physical well-being" were used as the preferred language in the influential Brambell Committee report in 1965. According to Gonyou:

> The report of the Brambell Committee more than any other document, identified ethology as relevant to the issues of the modern animal welfare movement. (p. 2172)

Of the five freedoms that were derived from the work of the Brambell Committee (e.g. Chap. 3), three were generated from agricultural concerns, while the last two (freedom to express normal behavior; freedom from fear and distress/avoidance of mental suffering) were seen as essentially ethological issues. Interestingly, the advance of applied ethology didn't bring many ethologists into the field of agriculture or veterinary medicine; it simply launched behavioral studies carried out by scientists untrained in the behavioral sciences. As animal husbandry and animal food production has developed, on farms, in biomedicine, and in zoos, behavior studies have been co-opted and specialists in ethology and psychology have been excluded. In our view this is a serious historical deficiency, but universities have begun to compensate by creating opportunities for trained animal behaviorists to work in veterinary settings and other applied programs.

To get the best results, zoos should attract the most qualified behavioral scientists to conduct animal welfare research, and operate programming units in applied ethology or behavior analysis. One idea generated from agriculture and germane to zoos is the notion that we should "fit farms (zoos) to animals, not animals to farms (zoos)". Applied ethologists have used a lot of the classic research paradigms developed by comparative psychologists, e.g. choice tests, schedules of reinforcement, so there is common ground for ethology and psychology wherever animals are managed by people. Bramblett (1989) effectively articulated the value of ethologists assigned to institutions that manage primates:

> . . . appoint an ethologist to the staff of each major primate facility or laboratory. Make the ethologist responsible for monitoring each animal on a regular, systematic and frequent schedule. In most cases ethologists are already at hand, but they should be explicitly integrated into all research with nonhuman primates: ethology should be just as fundamental a part of research design as nutrition or microbiology. This puts a professional in the role of advocate who can speak for the animals when necessary . . . The primary benefit of such a resource person to the institution is improved quality control. (p. 10)

Gonyou's call to recruit and teach applied ethology to agriculture and veterinary colleges will generate awareness about the need for ethology in animal management in breeding centers, laboratories, zoos and aquariums. A positive educational trend is the proliferation of interdisciplinary degrees such as the degree in animal behavior at the University of California at Davis. This campus has so many resources in animal science, veterinary medicine, and the behavioral sciences (including a national primate center) that it represents one of the best opportunities to obtain broad experience with a diversity of species (and a diversity of experts). Historically, it was one of the first American universities to support an exotic animal veterinary medical specialty.

5.3 Welfare for Elephants

The continuing controversy surrounding the welfare of captive elephants will be discussed throughout this book. Zoo animal welfare has become such a hot topic in the media due in part to public and professional concerns about the quality of life for elephants in captivity. Both circuses and zoos have been targeted by animal rights and animal welfare groups, some of which are clearly committed to the abolition of both. It is generally agreed that most circuses operate well below objective best practices in animal welfare, while zoos have an inconsistent history of elephant management. For example, an important study by Iosso et al. (2009) determined that circuses were far less capable than zoos in providing for the needs of elephants, big cats, and other popular circus animals. The high degree of stereotypy recorded for circus elephants is a clear indicator of poor welfare. Given the existence of a critical mass of substandard elephant facilities worldwide, zoos are collectively vulnerable to scrutiny and targeted criticism. The good news is that the best world zoos subject themselves to peer-review (accreditation) and periodic inspection by local, regional, and federal authorities, so there is ample opportunity for constructive criticism and improvement, and to differentiate accredited zoos from inferior roadside attractions. A significant trend to larger and better facilities in accredited zoos has been well documented in the mainstream media (e.g. Steele 2009; Chap. 7 this volume). However, optimal standards for elephants and many other creatures will always be debated (e.g. Maple et al. 2009). How much (or what kind of) space is necessary for an elephant or a whale to thrive? How many giraffe are optimal or acceptable as a cohesive group? Can a natural social system ever be established in captivity? Is training an enrichment tool or a constraint on the creature's autonomy and psychological well-being? Should the absence of stress be considered a deficiency?

The zoo or aquarium that is committed to optimal standards of animal welfare must keep standards and practices at the top of the operational agenda. Meetings with senior staff, curators, keepers, and at all levels of the zoo team, should never miss an opportunity to seek input about the current state of welfare in the zoo population. By encouraging discussion and debate, employees and volunteers will understand that animal welfare is indeed an institutional priority. It is not uncommon for front line caretakers to question the commitment of senior management in zoos and aquariums, so communication within the organization is very important. The timely application of contemporary research findings can be facilitated through modern computer technology. In the not-too-distant future every zoo and aquarium keeper will have a pc on their desk so they can keep in touch with the flow of scientific information, and keep up-to-date digital records on a daily basis. Curators, scientists, or veterinarians with access to the literature can create list-sharing opportunities to make sure the entire staff is following new trends and new ideas in design, enrichment, training, health and welfare. With the new information technology, caretakers have access to real time discussions with experts on a variety of subjects pertinent to innovations in zoo animal welfare. We need to do more to motivate our staff to generate new ideas and consistently support workable protocols currently in place. But there is no substitute for executive and board leadership to support an institutional commitment to the highest standards of zoo animal welfare (Fig. 5.2).

5.4 Primate Psychology

Nonhuman primates are the taxa that have received the greatest amount of sustained attention from scientists concerned with the issue of animal welfare. They have also benefited from a lengthy history of field, lab, and zoo research particularly in the specialty of behavior (Carpenter 1942; De Waal 1978; Harlow and Harlow 1965; Hinde and Spencer-Booth 1967; Kummer 1968; Lindburg 1969; Maple 1979b; Maslow 1936; Mason 1965; Washburn and DeVore 1961; Yerkes and Yerkes 1929). In a review produced by the Copenhagen Zoo, from a total of 200 environmental enrichment ideas, 87 applied to nonhuman primates, while 53 applied to carnivores, and 36 applied to birds (Griede 1992). Similarly, out of 20 articles on enrichment published in *Zoo Biology* from 1982 to 1991, 90 % concerned primates (Shepherdson 1993 as cited in King 1993). The history of the world's leading zoos documents remarkable achievements in the management and exhibition of monkeys and apes

Fig. 5.2 Crocodilians have successfully adapted to crowding in captivity

and represent useful animal models for other mammalian species. However, some welfare scientists have suggested that primate welfare is a bias that doesn't apply to the vast majority of species exhibited in zoos and aquariums. Vicky Melfi of the Paignton Zoo and Environmental Park in Devon, UK recently observed that the "taxa bias" led to an abundance of primate studies while many other species were rarely if ever the subject of welfare-oriented research. Further, Kagan and Veasey (2010) asserted that our greater focus on charismatic mega-vertebrates is counter- productive. There is a world of research potential in the zoo as we seek greater understanding of the needs of smaller, non-mammalian forms and begin to optimize living conditions for all. King's provocative essay presented an exhaustive list of reasons why avian species should be just as likely to be the subjects of enrichment research and enrichment protocols, asking:

> What are the criteria used for evaluating the need for environmental enrichment for a given taxon . . . Why does this bias toward mammals, and, in particular, primates and carnivores, exist? It is decidedly not due to their greater diversity or availability . . .

As King suggested, birds are at least as well if not better represented in zoo collections throughout the world, and we can certainly do much better in extending the reach of zoo animal welfare. If cognition is used as a differentiator, the propensity to use tools in nature has been extensively documented in many birds such as the Galapagos woodpecker finch, the Egyptian vulture, parrots, crows, and ravens (Beck 1980, Shumaker, Walkup, and Beck 2011). However, Beck has cautioned that tool behavior cannot be equated with intelligence as many animals, not just intelligent ones, have been observed making and using tools. Birds do experience unique welfare problems in the zoo environment. They are frequently the target of local nocturnal predators (e.g. raccoons, bobcats, and owls) that enter the zoo at night. Zoo managers also struggle with the problem that birds are often pinioned to prevent flight, a clear compromise in welfare. A less invasive procedure, feather trimming, still deprives birds of their principal mode of locomotion and restraint for trimming is stressful. On the other hand, structured bird shows enable controlled flight and this is a form of enrichment for raptors. With regard to exhibit design, birds have not benefitted from innovation when compared to mammals living in zoos, although there are some large flight cages for raptors living in rehabilitation centers throughout the country and in some major zoos.

Melfi (2009) also concluded that animal welfare's agricultural origins are not a perfect fit with the emerging specialty of zoo animal welfare. In her opinion, we must develop principles, standards, and practices that fit the requirements and opportunities that prevail in our specialized institutions. In a recently published book chapter, our research team (Maple et al. 2009) examined the criteria by which animals are judged as priorities for welfare. For example, animals with advanced cognitive skills were once regarded as a higher priority for cognitive enrichment, a prime example of "taxonomic elitism". We debated this notion and offered the following argument:

> Highly active species such as anteaters, bush dogs, capybaras, secretary birds, and zebras, most of which have not been tested by psychologists, also need room to roam and things to do . . . It is best to be egalitarian when it comes to saving or serving wildlife; big or small, bright or not so bright, beautiful or plain, each and every taxon deserves equal comfort and opportunity to express its species-appropriate character. (p. 134)

Controversies in elephant management are comparable to the issues faced by primatologists, veterinarians, curators and colony managers responsible for the history of significant change in primate exhibits and protocols. The primate model may extend to other species as well. Change did not come easily or quickly, but success in the primate realm was achieved largely due to the application of robust scientific findings from both field and laboratory observations and experiments (e.g. Mitchell 1970; Harlow 1971; Rogers and Davenport 1969; Dewsbury 1972; Davenport 1979; Erwin et al. 1979; Maple 1980; Maple and Hoff 1982; Bloomsmith 1989; Beck and Power 1988; Ogden et al. 1990; Maple 2007). The vast corpus of non-human primate research was extremely important in establishing the need for upgrades in exhibits and holding facilities, and zoo administrators and curators reached out to academic colleagues for advice and counsel in the exhibition and husbandry of nonhuman primates. Many workshops and conferences were organized throughout the world focused on relevant themes suggested and funded by seriously committed zoos and collaborating institutions of higher learning (e.g. deBoer 1979; Benirschke 1986; Norton et al. 1995). Smithsonian sponsored the 2003 elephant workshop that resulted in an important book, *Elephants and Ethics* (Wemmer and Christen 2008), and a recent workshop on the future of zoos was hosted by Canisius University. Videos from this meeting have been posted by the organizers (www.canisiusishar.org/symposia/futureofzoos.htm). Universities and university presses are incredibly important in disseminating new standards and innovations to benefit captive animals.

The growing public interest in elephants is driving the contentious debate about elephant welfare in captivity. However, it is more difficult to conduct experimental work on elephants so some of the assumptions behind reform have been fueled by speculation and anecdotes. On the other hand, field studies have provided a wealth of information about elephant behavior in nature, the necessary fundamentals for designing optimal elephant habitats in future zoos (e.g. Hancocks 2009). Michale E. Keeling

(Keeling et al. 1991), commenting on primate enrichment, once cautioned that we can be distracted in our zeal to exhaustively document our findings and, waiting for confirmation, fail to act when action is urgently needed. His views correspond to the ideas of Sommer (1974) who extolled the utility of his "action-research" approach. Action-research attempts to simultaneously manage and monitor change based on evidence. Both Sommer, an environmental psychologist and Keeling, a primate veterinarian, understood the compelling and urgent need to implement environmental reform. Action is required even if we don't have all the answers to our questions.

Because field workers have documented elephant home ranges over hundreds of miles (a mean of 880 km^2 for adult females in South Africa's Kruger National Park), and daily walks of up to eighty km (fifty miles), the most spacious zoo exhibits are regarded by critics as nowhere close to adequate. Private sanctuaries for elephants have been built on as much as 2,500 acres but this amount of space is also significantly smaller than a wild elephant's home range. It is difficult to imagine any zoo facility that favorably compares to nature in this dimension. Some critics have deduced that a failure to meet a natural norm is reason enough to prohibit the exhibition of elephants in zoos. There is no doubt that there are many restricted and inadequate elephant exhibits throughout the world, but the difference between zoo exhibits, sanctuaries, and the field cannot be measured by quantity alone.

The fundamental preference of quality over quantity is a principle first articulated by Hediger (1964) based on early field studies of birds (Howard 1920). Research on laboratory colonies of nonhuman primates has addressed space as an independent variable, confirming that quality of space is more important than quantity. In studies of macaques, Erwin (1979, p. 169) concluded that "social factors usually outweigh spatial factors." In a study of a large number of European zoos, Wilson (1982) discovered that the most important variables stimulating activity in great apes were "moveable objects" within the space. In a study of 29 orangutans in nine American zoos, Perkins (1992) confirmed Wilson's original findings. In designing Zoo Atlanta's gorilla exhibit, the design team was confident that the most important feature of the exhibit would be the large and contiguous social groups. Indeed, the animals began breeding immediately and formed cohesive social relationships. Remarkably, every gorilla but one born on-site has been raised by its mother from 1988 to the present. However, in a post occupancy evaluation (POE) of spatial behavior, Ogden et al. (1990) determined that the same gorillas at Zoo Atlanta did not optimize their use of space in one of the largest exhibits ever constructed in a zoo. Because gorillas in captivity that otherwise successfully interact and reproduce don't seem to need as much space as they utilize in nature, it is premature to generalize that elephants and other mega-fauna can only be successful in settings that approach the vastness of established home ranges in the wild. Optimal space may be best defined by qualitative rather than quantitative variables. Critical of the zoo to fulfill the needs of many species, Hancocks has observed that nonhuman primates need only sufficient social and cognitive challenges to provide for their welfare (Lemonick 2006).

Psychological well-being in the zoo can be regulated by many other factors including cover, enclosure furniture, verticality, variable feeding schedules, interactive technology, and social partners to name a few (e.g. Kortlandt 1960; Harcourt

1987; Maple and Hoff 1982; Maple and Perkins 1997). For elephants, rhinos and other animals that are active at night, a soft substrate and the opportunity to move around throughout the day and night may be engineered through design and training innovations rather than a commitment to massive enclosures. Two studies of how unrestrained zoo elephants behaved at night (Brockett et al. 1999; Wilson et al. 2006) demonstrated that elephants did not need to be tethered in their night-house. These findings contributed to the recognition that social contact with other elephants in night facilities is a superior management practice from a welfare perspective. Access to outdoor space at night is a promising best practice for elephants that is working for many progressive zoos. An extended day outdoors should work well for rhinos, hippos, lions and other species that are active at night in the wild.

Just a decade ago, animal rights activists advocated the "liberation" of zoo apes, but most apes exhibited in zoos today reside in spacious and complex simulations of the natural world, providing substantial social opportunity, spatial complexity, and a reasonable degree of autonomy. In the end, however, only systematic observation and controlled experimentation will provide the evidence to determine the right amount and the right kind of space for each and every species in the zoo. Husbandry standards should always be based on scientific inquiry rather than mere anecdotes or speculation. In the history of comparative psychology, a few species have been studied to determine the close connectivity of quality and quantity of space and behavior.

A highly relevant study by Nieuwenhuisen and de Waal (1982) bears on the issue of too many animals in too little space (social density). The authors of this report examined the response of chimpanzees (*Pan troglodytes*) to crowded conditions that prevailed when the animals at the Arnhem Zoo were moved indoors for the winter. A rodent model of crowding (e.g. Calhoun 1962) would have predicted chaos and social disruption; however Nieuwenhuisen and de Waal discovered that the big-brained chimpanzees essentially learned to cope. We know that human beings can cope with high social density so coping may be a function of an organism's cognitive capacity. If so, the conventional wisdom about elephant cognition would suggest that they too are capable of coping. An important question is whether translocations of elephants from one exhibit to another far distant location would lead to coping or severe emotional trauma. Currently, many field workers believe that disrupting the strong social bonds of elephants produces poor welfare. Moving elephants to fulfill demographic goals appears to be an example of putting the welfare of populations ahead of individuals. This is an issue that deserves careful consideration from elephant management committees worldwide. Planners should devote some attention to best standards and practices in translocations, and we should continue to study the behavior of animals when they are moved (e.g. Burks et al. 2004; Snyder et al. 2012).

Although western zoos rarely experience crowding, zoos in Southeast Asia are a different story. Agoramoorthy (2010) identified overcrowding as the most serious problem in Asian zoos, largely due to large numbers of confiscated and abandoned animals. It is ironic that the willingness of zoos to accept animals removed from deplorable living conditions (including circuses), has contributed to the malaise of poor hygiene associated with crowding at the zoo. As conditions worsen with each additional confiscation, Asian zoos are targeted for criticism by animal rights and animal welfare groups. In spite of invitations to collaborate with the zoos of the South-East Asian Zoo Association, none of their organized critics have stepped forward with any financial support to improve animal welfare. This is another missed opportunity to find common ground on behalf of animals.

Many primate experts agree that exhibits designed to challenge the animal's intellect and provide appropriate social stimulation can be classified as successful settings for apes. If apes thrive in the zoo environment, elephants and other big-brained mammals should also benefit from complex, naturalistic habitats, cognitive work stations, and social companionship. It will be important to factor this into our discussions of how much and what kind of space is necessary for a group of elephants, or any other social species, to achieve a satisfactory state of psychological well-being. With an animal as tall as an elephant, equipped with a trunk, verticality becomes an important dimension of space just as it is with an arboreal primate such as a gibbon, orangutan, or spider monkey. It is also important for climbing animals such as reticulated pythons. Verticality is essential for animals that nest in or explore high places (Fig. 5.5).

Abraham Maslow, Harry F. Harlow's first graduate student at the University of Wisconsin, studied dominance and sexual behavior in zoo monkeys long before he founded the field of humanistic psychology (Maslow 1935, 1936, 1940). In addition to his concept of a hierarchy of needs, which applies equally to humans and animals, he defined the inborn structure of psychological well-being that the senior author adapted for zoo animals in 1996 (Table 5.1 below). Essentially, this figure identifies the intrinsic propensity of primates to function in a socially positive way. We believe this represents the potential of nonhuman primates to achieve an optimal social life and therefore optimal wellness in a social context. By this model, the reader can see that it is the zoo keeper's job to encourage wellness through environmental and social engineering. Rather than study unhappy individuals or victims of psychopathology, Maslow chose to study high achievers and people who reached a high level of "self-actualization". His approach influenced the modern movement of "positive psychology" outlined by Seligman and Czikszentmihalyi (2000) who argued that psychology should concentrate on promoting mental health rather than its traditional singular focus on the treatment of mental illness. Surely a type of species-specific self-actualization in zoo animals is achievable in the Maslow-Seligman tradition, and it represents an opportunity to see if positive psychology can be broadly applied to a diversity of species in zoos and aquariums. This supports the proposition that investigators should use positive rather than only negative indicators of welfare (Chap. 3). A Maslovian-Seligman approach to welfare would identify successful animal exhibits and single them out for study. Maslow's early research in Madison was conducted at the Vilas Park Zoo where monkeys lived on an island and at the Central Park Zoo in New York. Because Harlow did not have a laboratory when he arrived as a new Ph.D. from Stanford, Maslow didn't participate in Harlow's pioneering studies of social deprivation, carried out with monkeys in small cages. Social groups in zoos provided the context to understand primate social behavior, but it was not conducive to experimentation. Maslow's exposure to group-living rhesus monkeys engaged in social behavior, dominance, aggression, and sexuality in a zoo

Table 5.1 Toward a comparative psychology of well-being (Maple 1996; adapted from Maslow 1962)

• The existence of a biologically based and uniquely individual species-specific inner nature.
• A nature that can be studied and discovered.
• A nature that is neutral, pre-moral, or good.
• Its nature should be encouraged; its suppression leads to sickness.
• A nature that is strong and unmistakable.
• A nature that is always pressing for actualization.
• Overcoming obstacles results in healthy self-esteem.

population, may have predisposed him to see human behavior in terms of its complexities and potential, rather than its limitations and liabilities. Therefore, Maslow's history of primate work may have influenced his ideas about human psychology more than most historians have acknowledged (Table 5.1).

5.5 Autonomy, Control, and Power

The significance of "autonomy" was articulated many years ago by Hal Markowitz (1978, 1979, 1982), an experimental psychologist with expertise in operant conditioning and environmental engineering. In his position as Research Director at the Metro Washington Park Zoo in Portland, Oregon (now the Oregon Zoo), Markowitz utilized operant conditioning techniques to train a variety of animals to work for food and the sheer joy of working. In every case described in his many publications, Markowitz devised an array of clever devices and situations to give the animals more control over their lives at the zoo. Asian elephants could pull a chain to produce a shower of water on demand; mandrills were encouraged to play games with visitors through the intervention of computerized touch-screens; and cognitive tasks were set up to produce food when the problem was solved. The animals were permitted to work at their own pace and on their preferred schedule. In the 80's and the 90's Markowitz experimented with more naturalistic technology in response to criticism of the "artificial" nature of his apparatus. His projects were a reaction to the barren nature of zoo habitats typical of this period. He continued to publish widely, influencing zoo design and husbandry throughout the world. One of his papers was the first publication on enrichment in the journal *Zoo Biology* (Foster-Turley and Markowitz 1982). Markowitz sought to create interactivity, novelty, and flexibility in the zoo environment, and provide animals with opportunities to exercise autonomy. His ingenious behavioral technology stimulated activity in even the most lethargic, solitary or deprived creatures. As Markowitz and Aday (1998) asserted:

> Nature is full of contingencies to which animals must learn to respond in effective ways. Where captive environments cannot include the replication of natural contingencies, un-natural ones may serve to provide animals with power. (p. 48)

After Markowitz was recruited to San Francisco State University to join its biology faculty, he began to improve the lives of animals at the San Francisco Zoo. His approach was much appreciated by the animal welfare community but welfare was not a word uttered comfortably by zoo professionals during the earlier stage of Markowitz' long career. Many of his original publications (e.g. in Erwin et al. 1979) branded his work as "behavioral engineering" but the terms were largely misunderstood. It was suggested that his use of technology reduced the animals to robotic automatons, but his critics missed the mark as Forthman-Quick (1984) convincingly argued. She noted that neither the naturalistic design approach nor behavioral engineering could solve every problem, but a combination of the two would be very effective indeed. Markowitz (2011) recently offered this explanation of his intent:

> Those who actually read or listened beyond the titles knew that we referred to engineering *environments* that would provide greater opportunities for resident animals . . . The shift to "enrichment" as the focal term . . . allows readers to better understand the intent of efforts in which we try to provide environments that give more power to others. (p. 4)

He introduced interactive devices into other zoos throughout the nation, notably at the Pana'ewa Rainforest Zoo and Gardens in Hilo, Hawaii. His ideas were embraced in Europe and prompted one official, UFAW Director Roger Ewbank, to make this interesting observation in 1986:

> Most European zoos lag far behind American ones in providing stimulation for their captive animals.

The occasion was a speech by Dr. Markowitz calling attention to his work at the London Zoo, and the announcement of UFAW's new Animal Welfare Zoo Award, for devices that enrich the environment of zoo animals. According to a contemporary press release (UFAW, July 18, 1986), Markowitz commented:

> Since most zoos are unwilling or unable to allow natural processes such as predation to occur on public exhibit, we must use contemporary engineering techniques to devise artificial contingencies to promote healthful exercise and maintain vigorous representatives of the species that we conserve in zoos.

Only recently have American zoos introduced animal welfare units that were sufficiently comprehensive and visionary to build on his legacy of creative intervention. In an innovation reminiscent of Markowitz, Steve Ross invented an aluminum hammer shaped like a potato and secured by a chain, to encourage chimpanzees at the Lincoln Park Zoo to break open macadamia nuts. Ross provided the animals with an anvil and demonstrated how to use it. Through imitation, hammering spread throughout the group and delivered a natural form of enrichment. As the man who inspired so much innovation in world zoos, Hal Markowitz should be acknowledged as one of the earliest pioneers of animal welfare in zoo settings. Later in his career he also contributed important ideas to advance the welfare of laboratory primates (e.g. Line et al. 1991). In his most recent book, Markowitz (2011) offered this suggestion to his followers (Fig. 5.3):

> We must focus on accepting guidance from those whose lives we wish to enrich, as well as on finding how to empower them to control more of their own lives. In the case of nonhuman animals this means that despite difficulties imposed by not being able to verbally communicate, we must make every effort to know as much as we can about the individual animals involved. Mastering knowledge about what works in getting others to perform in

Fig. 5.3 Hal Markowitz, the "father of environmental enrichment" (T. Maple)

the manner *you* wish is not the same as providing the opportunities for *them* to choose paths that are enriching. (p. 228)

Arnold Chamove was one of the pioneers of primate welfare with contributions that improved both zoo and laboratory animal environments (Chamove and Anderson 1989). His ideas on control are based on the behavior of animals in nature:

> In the wild, animals have more control over certain stimulus variables and most response variables than they do in captivity. Cages restrict animals' control, particularly with regard to their distance from stimuli outside and inside cages, as well as by limiting opportunities to act effectively on the environment. The discrepancy between control of stimulus and response variables in captive vs. natural environments, first elaborated by Hediger, may be the most fundamental issue in enrichment. (p. 191)

Nonhuman primates have been popular subjects for academic studies of both operant conditioning and cognitive psychology. Several decades of language research with great apes produced enclosure innovations by creative behavioral scientists such as Duane Rumbaugh who set up his first primate laboratory at the San Diego Zoo (e.g. Rumbaugh and Washburn 2003; Maple and Kuhar 2006) in 1958. Once Rumbaugh emigrated to the Yerkes Primate Center and Georgia State University, he continued to engineer unique social and structural opportunities for his experimental subjects. At the National Zoo, comparative psychologist Ben Beck developed a highly creative exhibit called "Think Tank" featuring language-trained orangutans that communicated with their keepers to earn food rewards and opportunities to utilize a series of arboreal pathways located high above zoo visitors and outside their enclosure (the "O" line). The largest primate that lives in trees, orangutans are rarely provided with sufficient arboreal opportunities in zoos. The National Zoo exhibit for this species is among the most innovative uses of vertical space ever devised for captive primates. First addressed in the enrichment literature and inspired by field research (Mackinnon 1974; Maple 1979, 1980), the vertical dimension of space is now regarded as a critical variable for the well-being of monkeys and apes in zoos. In virtually all monkeys and apes that climb, forage, play, and sleep exclusively in trees, it is preferable to think of space as essentially volumetric (Fig. 5.5). When volumetric space is made available to arboreal species, small exhibits become significantly larger and the usable portion of the exhibit is expanded. It is theoretically possible for a monkey or a bird to use every inch of a $20 \times 20 \times 20$ ft enclosure calculated at 8,000 cubic feet or more than 20 times the usable area of terrestrial space. This is a visionary concept not yet fully codified in accreditation standards or government regulations. One need only imagine the futility and frustration of a hapless gibbon or spider monkey without arboreal opportunities. We will return to the subject of vertical exhibitory in Chap. 8.

The O-Line is designed to encourage the natural locomotor propensities of orangutans, but Think Tank is essentially an exhibit of psychology and mentality. In fact, it is the only zoo exhibit that is entirely based on primate psychology and the scientists who have studied them. The intellectual power of primates is one of the primary drivers of design innovation as their psychological well-being depends on stimulating living conditions.

5.6 Social Organization

In the early 1970s there were still many adult male gorillas housed alone in zoos throughout the world. The more dominant living arrangement was a social system of enforced monogamy. Very few zoos in those days kept apes in groups. At the Atlanta Zoo in 1975, the primate house was comprised of one pair each of lemurs (*L. catta*), mandrills (*Mandrillus sphinx*), spot-nosed guenons (*Cercopithecus petaurista*), Debrazza's guenons (*C. neglectus*), and gibbons (*Hylobates lar*) in small enclosures, and one male lowland gorilla, one pair of common chimpanzees (*Pan troglodytes*), and a group of seven Sumatran orangutans (*Pongo pymaeus abelii*) in substantially larger cement, tile, and steel enclosures. The orangutans were moved to the zoo on loan from the Yerkes National Primate Research Center in an experiment on social living formulated by psychologist Richard K. Davenport (Maple 1980). Except for the group of orangutans, breeding was unsuccessful in this setting, and psychopathology was in evidence in every taxon exhibited there. Ironically, the only species in the primate house displayed in a complex social group was the normally solitary orangutan! At one point, the monogamous gibbon was housed with a closely related Siamang (*Symphalangus syndactylus*) and led to the production of a highly publicized hybrid. In those days zoos were more concerned with presentation than reproduction (Fig. 5.4).

In this era when many primates were prematurely taken from their mothers, zoo keepers and nursery volunteers began to compensate for the effects of isolation and social deprivation by carrying offspring in baby harnesses sold commercially as "snuglis". We regard this intervention as an early example of enrichment, even though we considered, then and now, that human hand-rearing of apes is a last resort (Maple 1980; Maple and Hoff 1982; Beck and Power 1988). Our best practices for great ape management stipulate that mothers should be encouraged to raise their own offspring, and managers should not be too quick to remove them.

Fig. 5.4 Snugli device enables keepers to serve as alloparents (T. Maple)

As zoos began to build more naturalistic exhibits they provided for species-appropriate social organization. To enable a naturalistic social network, all credible zoos today design exhibits for larger groups. Only the normally monogamous tamarins, marmosets, and gibbons live in pairs in zoos. This has been an extraordinary change and it has resulted in successful breeding, adequate parenting, and a blueprint for normal social development. Social composition in captivity should always be based on our knowledge of social organization in nature. However, the formation of social groups is tempered by the observations of Galindo et al. (2011):

> The idea that relationships with other individuals can be a major source of comfort and entertainment to captive animals is supported by evidence from a variety of species . . . On the other hand, damage that animals can inflict on each other by intimidation, overt aggression and redirected behaviours gives rise to serious animal welfare concerns. (pp. 228–229)

The competition to build bigger and better exhibits for popular animals such as gorillas has resulted in the exhibition of all-male bachelor groups, and bachelor groups represent some risk of aggression and injury. While it is still difficult to obtain a family of gorillas, surplus males are available. Bachelor groups of gorillas occur in nature, so zoo biologists are studying them to see if it is feasible to expand the practice (Stoinski et al. 2004). The Lincoln Park Zoo (Chicago) recently introduced an all-male group with considerable fanfare, the result of a decision by the SSP gorilla committee.

The organized effort by accredited zoos to establish species-appropriate social groups of nonhuman primates is a major achievement. This advance led to the growth of healthy self-sustaining populations through strategically planned and organized management committees working as a group to acquire, monitor, and utilize demographic and behavioral information. Further, complex social groups provide a more accurate illustration of how a species lives in nature. However, the more we study how animals live in the wild, new information leads to changes in our captive management policies. Now that we know that wild gorillas sometimes tolerate more than one silverback male in the group, we are experimenting with this idea in the zoo. If the practice became more common, it would reduce the number of surplus males. An isolated male is a welfare problem, while all-male groups do present some welfare risks. We don't yet know if an additional adult male will lead to serious fighting. The unpredictable nature of group dynamics is another good reason to seriously monitor social groups on a daily basis, and justification for a formalized research program. A case in point is the recent death of an infant chimpanzee killed by the dominant male of a very large social group at the Los Angeles Zoo.

Brenda McCowan and her associates (2008) at the California National Primate Research Center recently conducted research to determine the utility of "social network analysis" to decipher patterns of aggression and wounding in rhesus monkeys. The study collected 37,000 observations of affiliative, submissive, and aggressive behaviors for 1,300 macaques over a three year period. The results of this study suggested that social power, displacement fragmentation, and grooming reciprocity within groups were significantly associated with rates of contact aggression and wounding, and the occurrence of severe aggression known as "cage wars". The investigators also determined that manipulations of group composition and matriline configurations promote social cohesion and stability in social groups. As social network indicators could be valuable predictors of instability, this tool may be useful to managers of other primate species in zoos. McCowan's research is a good reason to keep informed about relevant laboratory-based studies. Twenty years ago, scientists from primate centers and zoos collaborated and held joint symposia at national meetings of scientific societies such as the American Society of Primatologists, the Animal Behavior Society, and the American Association for the Advancement of Science. Other productive collaborations between zoo biologists and environmental psychologists led to joint publications, competitive research grants, and successful programming of award-winning zoo exhibits and educational innovations. Regrettably, collaborations of this quality are few and far between today. For an exception, see the published symposium organized by Koch (2007).

The recent cross-disciplinary review by Galindo et al. (2011) drew some important conclusions about social issues and animal welfare:

1. Many animal welfare problems in social groups are a result of not taking into account the social history and evolved social capabilities of the group members.
2. Attempts to solve social problems in captivity have to focus on a deep understanding of social behavior, including the role of individual differences within a dynamic social environment.
3. Although the social environment, if badly managed, can be a constant source of trouble, it is also a potential tool to improve the welfare of individuals, as it is the most important source of stimulation, interest and comfort in the lives of many captive animals (p. 241).

Fig. 5.5 Orangutan climbs to height of 54 ft at Zoo Atlanta (T. Maple)

5.7 Personality and Welfare

Personality can no longer be ignored as a factor in zoo management. Personality is a psychological construct that has been demonstrated in a diversity of species, e.g. baboons, cats, chimpanzees, cichlids, dogs, goats, gorillas, hyenas, macaques, octopuses, squirrel monkeys, and wolves (Gosling 1998). Recent research, reviewed in the following paragraph, puts elephants into this category for the first time. A better understanding of personality variables might help veterinarians, curators, and keepers respond to idiosyncratic behavior patterns, particularly issues related to aggression and compatibility. Habitat variables, such as cover and enclosure furnishings, might serve to reduce aggression or moderate the severity of its expression.

As a graduate student Ken Gold carefully studied the entire North American population of lowland gorillas. A travel grant enabled him to observe many of them first-hand with the cooperation of zoo curators and keepers. His intuitive belief that personality and temperament were factors in successful introductions led him to carry out a doctoral study of personality in zoo gorillas and create an instrument known as the Gorilla Behavior Index or GBI (Gold and Maple 1994). His extensive correspondence with a nation of zoo keepers provided personality assessments for 298 gorillas in North America. Theoretically, he suggested, SSP managers could utilize this information in predicting the success of translocations. The information could also be used to evaluate special behavioral needs of animals when groups are assembled in a zoo that is undergoing renovation. The technique of personality assessment through factor analysis is well understood and now extended to a diversity of creatures that live in zoos. There is no good reason why we cannot generate personality profiles to be used in managing them. If psychological well-being requires the opportunity to express a full range of emotions, certain animals

Table 5.2 Assessments for behavioral traits (Adapted from Highfill et al. 2010)

Factor	Trait	Assessment (test)
Openness	Curious	Open field test
	Not exploratory	Open field test
Conscientiousness	Perseverant	Unsolvable puzzle feeder
	Careful, cautious	Latency
Extroversion	Active, energetic	Observations
	Timid	Open field reactions
Agreeableness	Affiliative	Y-maze
	Friendly	Reactions to people
Neuroticism	Self-satisfied, free from anxiety	Observations
Aggressive	Dyadic	Observations in the presence of puzzle feeder

will require extraordinary facilities to meet this objective. This is an aspect of zoo animal welfare that has not been fully addressed in the literature. Kuhar et al. (2005) later utilized the GBI to compare data to an ongoing study of all-male groups. It is anticipated that personality data used together with other information will aid in predicting cohesion in newly established social groups of gorillas. These techniques should be useful in the formation of other group-living nonhuman primates where morbidity and mortality is a concern, including baboons, chimpanzees, guenons, langurs, macaques, and prosimians (Table 5.2).

A forthcoming article in the journal *Applied Animal Behaviour Science* (Grand et al., 2012) used personality ratings to determine components of personality in captive African elephants. The investigators used a modified version of the Gorilla Behavior Index developed by Gold and Maple (1994), resulting in an instrument labeled the Elephant Behavior Index. Four personality components were identified from ratings by 16 observers familiar with the animals; (1) Fearful; (2) Effective; (3) Sociable; (4) Aggressive. These factors correlated with salivary and blood serum cortisol. Morning cortisol levels were positively correlated with the fearful component, while the other factors (effective, sociable, and aggressive) were negatively correlated with morning cortisol levels. These measures were identified for use in refining management protocols for individuals as elephants pose a high risk to keepers.

Bull elephants are so dangerous in captivity that they require barriers that are essentially indestructible. When the great size and strength of a male elephant is combined with temperamental fluctuations associated with the hormonal surge of musth, holding facilities will be put to the test. Extraordinary precautions must also be taken when animals such as elephants are bred, as breeding can be a very violent albeit brief encounter. Big cats and primates can also suffer serious injuries when paired for breeding, so holding facilities must be designed with this purpose in mind. Breeding decisions are still influenced by genetic considerations rather than compatibility, and careful introductions are always necessary. Valuable studies on introductions have been conducted by Burks et al. (2004).

Although many species have been subjected to personality assessments, the vast majority of animals remain a mystery in this regard. By renewing the interest of psychologists, the vast population of zoo and aquarium animals represents an unprecedented opportunity to extend the reach of personality research, one of the most important fields of psychological inquiry. A unified theory of personality, constructed from research on closely related primates and other advanced mammalian forms, may be possible in the near future.

5.8 Psychopathology in the Zoo

The exhaustive research program of Harry F. Harlow and his students and collaborators at the University of Wisconsin (e.g. Harlow and Harlow, 1965; Kaufman and Rosenblum 1967; Mason 1965) provided a wealth of information that can be used to guide management decisions in zoos. Isolation and social deprivation was widespread in the 1960s when monkeys and apes were imported in great numbers by laboratories, primate centers, and zoos. It was at that time standard practice for these institutions to remove infants from their mothers soon after birth. This practice reveals a lack of understanding about the norms of parenting, and the deleterious side-effects of social isolation. As Harlow clearly demonstrated, "motherless" monkey mothers repeated and extended the syndrome by ignoring or neglecting their own offspring. Comparative psychologists also documented long-term cognitive deficits in chimpanzees due to restricted rearing early in life (Davenport et al. 1973). Zoo managers have learned from this large volume of psychological research, conducted mainly in primate centers. Data from Species Survival Plan queries are sorting out the trends. According to Beck and Power's (1988) survey of North American institutions, only one-third of captive-born gorillas were mother-reared at the time of his research. When Hoff et al. (2005) examined the data for 1995 gorilla births in U.S. zoos the trend had reversed with 12 out of 18 gorillas mother-reared.

Although Hediger (1950) and other European ethologists documented the phenomenon of stereotypy in zoo animals many years ago (Meyer-Holzapfel 1968), environmentally-induced psychopathology in monkeys and apes was the norm in zoos until enlightened curators and colony supervisors began to manage their collections in a different way. Today, there is an unwavering commitment by keepers, curators, and in-house behavioral scientists to prevent psychopathology by enrichment methods, naturalistic housing, the maintenance of appropriate social groups, and the encouragement of mother-rearing. Fifty years of psychological research provided the intellectual framework to facilitate normal socialization in zoo populations of primates (e.g. Harlow and Harlow 1965; Davenport 1979; Erwin et al. 1979; Beck and Power 1988). Primatologists working in the field also provided context for some unusual behaviors such as coprophagy (consumption of feces). Although it is rare in the wild, Harcout and Stewart (1978) documented this behavior in wild gorillas confined to their nests during or soon after rainy weather. Interestingly, one of three explanations for this behavior offered by the investigators was boredom. In captivity, coprophagy is an unsavory habit that disturbs visitors.

Abnormal behavior is thought to interfere with the zoo's positive messaging but until recently there wasn't any evidence to support this contention. In an experimental study, Miller (2012) played a short video of tigers pacing or at rest and asked human subjects to rate the level of animal care. After observing a pacing tiger, respondents gave the institution a low rating on animal care, and indicated less interest in supporting the zoo. As Miller noted, customer loyalty is important to any organization and repeat business can be affected by the impression gained at the zoo. Abnormal behavior, including stereotyped pacing leads to visitors feeling sorry for the animals and blaming the institution for their plight. Perceptions of poor welfare should be a warning that intervention is necessary for the good of the animal and the zoo's reputation. Too often animals arrive at our institutions with a history of compromised welfare. In such cases, deprivation acts or stereotypies may persist in facilities that are objectively optimal.

Zoo animals that suffer evident signs of fear and anxiety should be provided behavioral or medical treatment. A product developed for dogs and cats has proved to be effective in reducing fear and aggression during thunder storms, a regular occurrence that terrifies many pets. The "Thunder Shirt" simulates the close comfort provided by an owner who holds the pet. The product is based on the idea advanced by animal welfare expert Temple Grandin that pressure has a calming effect on the nervous system. For example, veterinarians have used pressure to calm cattle during vaccination. The shirt, based on agricultural restraint devices, also seems to help when dogs and cats are examined by a veterinarian. A similar product is available under the brand name AnxietyWrap. As Harlow demonstrated, cloth surrogates are a very effective substitute for a mother's love when rhesus monkeys are raised in isolation. The same principle seems to be operating in this case. Perhaps a zoo version of the thunder shirt could provide comfort on those occasions when an animal cannot be consoled in any other way. Roger Caras (1995) articulated his concern with fear in his essay entitled "A View from the ASPCA" (Fig. 5.6):

> . . . the purpose of pain is to tell us when and where something is wrong . . .
>
> I would add that animals in the wild and in captivity are frequently afraid. That is a survival mechanism. Some practices in zoos can unknowingly, and with the best of intentions, impose fear and anxiety on animals. It may be projection on my part, but excessive, unwarranted, unrelenting apprehension to me is a cause of pain, an element of cruelty in the extreme. (pp. 298–299)

Bradshaw and Sapolsky (2006) have proposed a new approach for dealing with the effects of confinement. They combined psychology, ethology, and neurobiology into a cross-species model of brain and behavior and labeled it "trans-species science". The authors utilized this approach in comparing humans and elephants in confinement. They argued that the construct "Complex Post-Traumatic Stress Disorder" explains behavior in both elephants and humans suffering the stress of captivity (Bradshaw 2009). Bradshaw traces the syndrome to serial trauma; 1. The cull-capture experience; 2. Premature weaning and compromised rearing; 3. Loss of primary attachments by enforced separation; 4. Transport and translocation; 5. Environmental deprivation. Bradshaw identifies captive elephants as candidates for full-blown PTSD. His remedy

Fig. 5.6 Isolate rhesus monkey exhibiting a deprivation act (B. Redican)

is to cease the capture and culling of wild elephants, a cessation of captive breeding programs that break up relationships, and an end to elephants living in "close-confinement captivity." As compelling as these ideas may be, it should be noted that Bradshaw's paradigm is not empirical in its formulation, and the comparison of elephants directly to humans is a significant departure from the normal conventions of comparative psychology.

5.9 The Behavioral Basis of Design

In his classic book *Tight Spaces* (1974), environmental psychologist Robert Sommer observed: "The hard zoo consisting of concrete boxes, steel bars, and fixed routine of feeding, watering, and washing by outside maintenance personnel, distorts the behavior of animals". The educational message of hard zoos portrays the captive animal as a virtual felon, a creature to be feared, requiring distance, restraint, and punitive methods of control. As we have learned, the stereotype of "King Kong" or the circus ape "Gargantua" is not consistent with fieldwork on the gorilla. A primate of enormous size and strength, capable of defending the group from virtually any predator or intruder, the silverback male gorilla is generally a peaceable creature in nature. As we have learned, gorillas can be habituated to visitors who trek long distances just to get close to a wild gorilla. When designers started to plan gorilla exhibits to fit the data rather than the myth, species- appropriate social groups could be established in zoos.

Normal gorilla behavior requires a process of socialization. To establish species-appropriate social groups in captivity, we depend upon data gathered by field biologists. The entire program for Seattle's Woodland Park Zoo gorilla exhibit was based on knowledge gained from field studies in Africa. The immersion concept pioneered in Seattle was the basis for improved facilities in Atlanta (Ogden et al. 1990), San Diego (Ogden et al. 1993, 1994) and New York (Conway 1999). Each successive exhibit was designed to improve upon the innovations of the others; for example, the Atlanta exhibit provided for a population of gorillas separated into five contiguous naturalistic enclosures. For good reason, charismatic gorilla exhibits piqued the interest of decision-makers and designers, and for many years attracted gifts big enough to drive innovation, imitation, and expansion. Fortunately the features of these exhibits have trickled down to inspire improvements for other primate taxa, but the incredible diversity of the primate order is under-utilized in most zoos. In terms of design, it appears that gorillas are the flagship species for the primate order. Given all that we have learned about gorillas in the past thirty years, it is not surprising that so many superior exhibits have resulted but exhibits continue to evolve and some of the best exhibits of the past 30 years are now ready to be updated and renewed (e.g. Fig. 5.7).

Clearly, the foundation of superior design is behavior, and wellness/welfare issues loom large as design priorities. Architects don't generally test their assumptions, but psychologists do. For example, Jon Coe's appealing and logical proposition that animals should be exhibited in a position slightly higher than the visitor to inspire and generate respect has never been evaluated. Similarly, animals presented below grade are thought to generate derision or abusive behavior from visitors, a clear lack of respect. This too is an attractive hypothesis, but we have no evidence to support the claim. Jon Coe (1996) summarized his intuitive design philosophy in the following paragraph:

> Design the exhibit in such ways that the animals appear to dominate the scene. Always present the animals respectfully if you expect visitors to treat them respectfully. Try to avoid creating situations in which the public looks down on the animals. If possible, place the animals higher than the visitors. (p. 171)

To establish the connection between design and welfare, we should gather evidence through post-occupancy evaluations of new exhibits. Scientists have examined the effect of exhibit changes for a variety of species in the zoo (Maple and Finlay 1986; Ogden et al. 1990; Ross and Lukas 2006), and many studies have included zoo visitors in the scope of the survey (Finlay et al. 1988). From this information zoo biologists are building a case for future zoo design, another salient dimension of the empirical zoo. We will have more to say about the science driving the new trend of "activity-based design" in Chap. 8. Post-occupancy evaluations can also reveal how animals adjust to the disruptive effects of construction. Animals must often endure the chaos of work crews and high frequency, high decibel noise of their equipment and technology. In urban settings, there is no escaping the noise that surrounds the zoo campus, when the noise of mass transportation, police and fire vehicles, and other industrial acoustical stimuli are everyday occurrences. In a study of two giant pandas exposed to demolition

Fig. 5.7 San Francisco Zoo gorilla habitat c. 1980 (T. Maple)

work at the National Zoo (Powell et al. 2006), both animals exhibited acute behavioral and endocrine responses, but the investigators concluded that the noise of demolition did not significantly reduce welfare. These results are supported by an earlier study at the San Diego Zoo in which visitor noise produced minor responding but did not affect welfare (Owen et al. 2004).

A clear indication of the power of psychology as an intellectual foundation for modern zoo design is revealed in the words of two experienced zoo professionals, John Seidensticker and James Doherty (1996) (Fig. 5.7):

> When someone makes a decision to visit the zoo, that person is usually already highly motivated. Once at the zoo, the visitor's expectations are strongly influenced by a hierarchy of needs: physiological needs, safety needs, social needs, and needs for self-esteem, self-actualization, and creativity. Unpleasant encounters with excessive trash, dirty bathrooms, poor directional signs, and failure to see animals that were expected can lead to great disappointment, or what can be termed 'cognitive dissonance'; a situation in which a person experiences something that was neither expected nor desired . . . Curators have a very important stake in encouraging excellence in those programs that strongly influence the visitor experience. (p. 183)

The foregoing observations by zoo biologists mirror the interests of academic environmental psychologists who have created a massive body of information about how people respond to the natural and the built environment. A good starting point for a zoo designer is the book *The Environment and Social Behavior* by Irwin

Altman (1975). This source like many early books on environmental psychology contains classic contributions to our understanding of human spatial behavior and its antecedents in the literature of animal behavior and ecology. There is an impressive literature in the field of "environment and behavior" and the data from human studies can actually be applied to animal problems that we study in the zoo, e.g. territoriality, personal space, privacy, conflict, crowding and density, coping, and nonverbal communication.

As we have repeatedly emphasized there are great benefits when zoos and aquariums collaborate with colleagues in academia. A very good example of the role of psychology is a field study by Bates and her associates (2007) who demonstrated that African elephants were able to classify human ethnic groups by their odor and garment color. This elegant field experiment was guided by Dr. Richard W. Byrne, a professor of psychology at the University of St. Andrews in Scotland, and carried out at the Amboseli field site coordinated by Cynthia Moss and Joyce Poole. The investigators interpreted their findings from the perspective of cognitive ethology:

> Elephants therefore show remarkable discriminatory abilities: the ability to use olfaction *and* vision, independently, to classify garments according to their likely human wearers, and to vary their reactions appropriately to the probable danger. Given the potential adaptive benefits of classifying a predator species into subcategories, we expect that this ability will prove to be widespread among animals with appropriate perceptual and cognitive capacities. (p. 1–5)

Chapter 6
Environmental Enrichment

We can, as keepers of the captive biota, give the pleasure of intelligent, appropriate companionship to sentient beings, captive or otherwise, raising life to higher levels of appreciation for both parties. We can develop such bonds knowingly, with purpose, and use such in management or reintroductions. Ethological studies teach us that life in captivity must not be a prison sentence for wildlife.

Valerius Geist

Enrichment is potentially one of the most powerful tools an animal caretaker has to improve welfare for an individual. Environmental enrichment has been conceptualized in various ways, but the definitions generally relate to adding sensory stimuli or providing choices in the environment (de Azevedo et al. 2007) in an effort to increase behavioral opportunities to benefit the inhabitants. Viktor and Annie Reinhardt (1998) defined environmental enrichment as "the provision of stimuli which promote the expression of species-appropriate behavioral and mental activities in an understimulating environment." Young (2003) offers the approach of defining the goals of environmental enrichment rather than the concept itself. He modifies the descriptions of Shepherdson (1989) and Chamove and Moodie (1990) to propose the following goals for enrichment: "(1) Increase behavioral diversity; (2) Reduce the frequencies of abnormal behavior; (3) Increase the range of normal (i.e., wild) behavior patterns; (4) Increase positive utilization of the environment; (5) Increase the ability to cope with challenges in a more normal way (p. 2).

By focusing on these goals, we can develop effective and efficient management practices for zoo animals. In many cases, enrichment has been used in a posthoc effort to reduce stereotypic behavior or correct other deficiencies, such as inadequate exhibit design, but it is being used increasingly more often in zoos in a proactive manner to create a rich, stimulating environment. Enrichment is covered in virtually all husbandry plans issued by regional and national associations, so managers and keepers are aware of the options. However, enrichment is not always successful (Mason et al. 2007) and there is a need for critical assessment of what types of

T.L. Maple and B.M. Perdue, *Zoo Animal Welfare*, Animal Welfare,
DOI 10.1007/978-3-642-35955-2_6, © Springer-Verlag Berlin Heidelberg 2013

Fig. 6.1 Gorilla enjoying cover in artificial tree at Zoo Atlanta (A. Clay)

enrichment work, with what species, in what contexts, and how to implement these practices in an effective and cost-effective manner. This chapter will outline some of the existing frameworks for defining types of enrichment and then touch on the critical issues that should be considered as the use of enrichment grows (Fig. 6.1).

Enrichment in the United States became a priority after the passage of the 1985 Animal Welfare Act introducing the concept of "psychological well-being" (Adams 2007). The legislation was particularly pertinent to non-human primates and dogs managed in research settings. An equal emphasis on enrichment and exercise is an important dimension of the legislation as exercise is so essential in managing health and wellness (Chap. 4). Legislation worldwide continues to prioritize enrichment procedures to benefit animals in zoos and aquariums. In this chapter we review the full scope of enrichment as it is currently practiced with a diversity of captive animals.

6.1 Types of Enrichment

In a survey of zoo institutions, Hoy and colleagues (2010) described eight types of enrichment: feeding, tactile, structural, auditory, olfactory, visual, social, and human-animal. Their review is comprehensive; we suggest only the addition of cognitive enrichment to create a complete framework for identifying relevant categories of enrichment. We will briefly consider each of these categories.

6.1.1 Feeding Enrichment

This type of enrichment requires manipulation of the food itself or the means through which it is delivered. Rather than providing all of an animal's daily rations in an easily accessible, single delivery, zoos have shifted towards practices such as scatter feeding (spreading food around to require travel between locations to obtain all items), devices that require manipulation to extract the food, or required behaviors to obtain food. This begins to more closely represent conditions in the wild in which an animal would spend a great deal of time foraging or hunting for food. In an interesting study investigating stereotypic pacing in tigers, Jenny and Schmid (2002) installed electronically locked feeders in the enclosure. At random times, the boxes unlocked and food could be obtained. The authors observed a significant decrease in pacing, compared to the standard situation in which all food was provided at once. The animals had more control over the delivery of food (by consistently checking boxes) than when food was delivered irrespective of the animal's behavior which may have contributed to the changes in behavior. Cummings and colleagues (2007) found food enrichment to be more effective than object enrichment at increasing exploratory behavior in Maned wolves (*Chrysocyon brachyurus*). When four self-operated food boxes were introduced to a rainforest exhibit at the Zurich Zoo, lemurs (*E. fulvus*, *H. griseus*, *V. variegata* spp.) were more active especially in trees during feeding times and the animals engaged in more natural behaviors in the presence of zoo visitors. Prior to the introduction of elevated feeders, the animals spent their time on the ground, but the feeders encouraged them to become arboreal. Zoo keepers reported that the animals quickly learned how to operate the feeders (Sommerfeld et al. 2006).

Kuczaj et al. (1998) described feeding practices at Sea World where orcas (*orcinus orca*) are fed on a variable-ratio schedule (defined as reinforcement delivered after an unpredictable number of responses) in order to eliminate habituation or expectancy. This enabled animal managers and trainers to keep orcas in an unpredictable environment. In a typical day, the enrichment agenda for the orcas at Sea World included exercise, learning, show, husbandry and veterinary care, research, and play. Sea World, and other marine parks, traditionally engaged in cognitive research with orcas and other marine mammals, and regard research interaction as a form of enrichment since the animals are eager to participate. The value of academic collaboration is demonstrated by the high productivity of Louis Herman and his colleagues (1987) who have worked with Sea World personnel to investigate the learning abilities and sensory modalities of marine mammals for decades. Professor Herman brings resources from the University of Hawaii system through his appointments in Psychology and Oceanography. He has published more than 120 scientific papers on communication, perception, and language. For complex marine mammals such as dolphins and whales, enrichment needs to be a constant and daily form of interaction. The demands of an active research program will accelerate the pace for animals that benefit from intellectual challenges (Fig. 6.2).

6.1.2 Tactile Enrichment

This type of enrichment involves the provisioning of objects such as bags or balls that are physically stimulating to an animal. This can involve relatively simplistic

Fig. 6.2 Komodo dragons are excellent subjects for target training (A. Thompson)

manipulations to the environment such as spreading hay ranging to more complex manipulations such as installing floor panels that provide extra heat or cooling when reclining. Baker (1997) found a reduction in stereotypic behavior in chimpanzees when the animals were provisioned with straw bedding. Manipulations of this kind are cost-effective and easy to implement. In an assessment of deep litter, Fuller et al. (2010) discovered that Wolf's guenons (*Cercopithecus wolfi*) were more active and spent more time feeding when they were presented with either wood wool (excelsior) or straw litter, but straw was more effective than wood wool products. However, the use of deep litter didn't affect agonistic behavior, one of the goals of the intervention.

Water can be a significant source of enrichment for many animals. Bears and some great cats need water features so they can swim. Jaguars (*Panthera onca*) at the Palm Beach Zoo frequently entered their shallow pool to retrieve live fish when presented or to play with plants. A mother often played with her offspring in the water as it seemed to excite them. Tigers are also fond of water in the wild and in captivity. They are good swimmers, but breeding tigers with offspring risk the loss of their cubs if their pools are too deep, prompting keepers to drain pools until the cubs have matured. Lowland gorillas in the Central African nation of Congo enter swamps to locate delectable food items, but gorillas apparently cannot swim and are prone to drowning (Maple and Hoff 1982). Research by Susan Brown and her colleagues (1982) found that a male gorilla's entry into water at the Audubon Zoo

Fig. 6.3 Lowland gorillas will enter shallow water to soak and to play (T. Maple)

was influenced by humidity but not by attendance or by ambient temperature (Fig. 6.3). The investigators concluded that shallow water provided environmental enrichment for lowland gorillas in this setting. Elephants enjoy pools but they are even more enamored of mud wallows. Both water and mud generate a lot of excitement from elephants. A mud wallow is enriching for rhinos and warthogs as well. A comfortable environment for many animals requires a full immersion pool, mud, and plenty of available dust. Hediger (1964) observed, however, that while many species of hoofstock like to wallow, there are others that show no such inclination. One drawback to water is the increasing cost of providing water for horticulture and other purposes at zoos and aquariums. In many parts of the world, water is a diminishing resource so this enrichment strategy may not be sustainable in the long term. Without access to water features the exhibition of capybaras, crocodiles, hippos, manatees, sea lions, tapirs, waterfowl and dozens of other water-dependent species would be impossible (Fig. 6.4).

6.1.3 Structural Enrichment

Structural changes usually refer to longer-term or semi-permanent changes to an animal's environment such as the introduction of a new platform for sitting or ropes for swinging. At Zoo Atlanta, a large, artificial tree was built in the orangutan exhibit. The tree contained several enrichment devices such as a computer for cognitive testing with a pellet-delivery system, water sprayers, and a heated platform. Researchers found that despite the potentially high value of this resource, there were no significant changes in aggression, stereotypic, or distress-related behaviors when comparing conditions in which the tree was activated or not (Perdue et al. 2012). For arboreal species, it is difficult to accurately simulate rainforest climbing opportunities (Fig. 6.5). Trees made of cement or gunite are inflexible and turn the exhibit into a literally hard architectural environment. Living trees must be protected by hot-wired or

Fig. 6.4 Warthog enrichment by access to a mud wallow (B. Perdue)

chain link wrappings so the animals cannot damage them by climbing and feeding on the branches, or remove the surface bark. In a contest between the well-being of wildlife, or the survival of horticulture, the trees usually win. Thus, living trees with any size provide nothing more than shade. One compromise provides for lower story replaceable plantings that are available for foraging and nesting purposes, along with the effective protection of mature trees. Artificial climbing structures can be added to the mix to promote movement in vertical space. Many zoos have experimented with flexible poles that function and look like a bamboo forest. Gibbons and orangutans are adept at utilizing these functional but artificial trees for arboreal locomotion and to engage in aggressive displays (Fig. 6.5). A perturbed orangutan will severely test the strength of innovations in exhibit technology.

Frediani (2009) examined the potential for plant based enrichment in naturalistic zoo habitats. He observed that plants have been overlooked and not utilized to their full potential as sources of environmental enrichment. Plants can be utilized as a form of social, cognitive, physical, sensory, and food enrichment. For example, a living tree has advantages over a fixed climbing structure; it is regenerative rather than degenerative, and it provides shade, fruit, and flowers. Large trees have survived in zoo enclosures with many species, even nonhuman primates as large as apes. Gorillas entered enclosures with trees in Seattle and in San Francisco soon after the animals were introduced in the 1970s and the trees still survive to this day. Frediani's paper invites the reader to examine the BIAZA plant wiki (Seiffert 2009) to discover new ways to utilize living plants as enrichment. Most zoos provide browse for enrichment and as food supplements. For example, at the San Francisco Zoo, where Eucalyptus trees grow in abundance on its campus, gardeners collect browse for koalas (*Phascolarctos cinereus*) 5 days each week. The horticulture team devotes 54 h per week to harvest browse, amounting to 61 tons of material annually (Beach, personal communication). Zoos with specialized consumers, such as koalas and giant pandas, often resort to planting *Eucalyptus* or bamboo so they can harvest at the zoo on a regular basis.

6.1.4 Auditory, Olfactory, Visual Enrichment

Providing species-appropriate sounds may have the effect of enriching an animal's environment. This type of enrichment will require curatorial knowledge about which sounds are appropriate for a given species. For example, playing lion roars to a group of African hoofstock might be frightening. On the other hand, recorded lion roars played to other lions may be enriching to the lions, as well as to zoo visitors (Kelling et al. 2012). In fact, Kelling's study enhanced the visitor experience as measured by exhibit stay time and the behavior of zoo visitors. In this study, the playbacks of lion roars did not increase visible fear in nearby ungulates, nor did it adversely affect other lions, but naturalistic playbacks did increase live roaring. Lion vocalizations in this instance can be regarded as an effective form of auditory enrichment. The vocalizations of other great cats should be investigated to see if they are benign in their effect on other animals. In gorillas, auditory enrichment, from authentic forest sounds of birds and insects, has been found to decrease stress-related behavior and marginally affect abnormal behavior (e.g., Ogden et al. 1994; Wells and Irwin 2008). Markowitz et al. (1995) introduced a hunting game to a 16 year old African leopard (*Panthera pardus*) in which bird songs were played and pursued. The subject had spent most of its life in a sterile cage. The pursuit activity produced a food reward and affected indicators of psychological well-being, increasing activity and reducing stereotyped behavior (Fig. 6.6).

In a very clever study by Wells and Irwin (2008), four female Asian elephants (*Elephas maximus*) were exposed to two conditions of auditory stimulation; the experimental condition provided audio stimulation from a commercially-available CD of classical music. Stereotyped behavior was reduced during the experimental condition when compared to the control condition. This form of enrichment in elephants is promising. The well-documented ability of Asian and African elephants to hear low-frequency sounds may present other experimental opportunities for the study of auditory stimuli as enrichment (Payne et al. 1986).

Olfaction is an incredibly important sensory mode for many animals, and the zoo provides a great setting in which to capitalize upon the rich array of smells readily available everywhere. Odors from other individuals, even other species, can be spread in an animal's enclosure through urine, feces, or materials that promote interaction with the substance. Of course, this must be done within zoo safety and health regulations, but there are many great opportunities that can be exploited in this context. Furthermore, many animals respond to manmade odors (cinnamon, vanilla, mint) and smells such as Tabasco. This also provides a joyful opportunity for zoo visitors to watch monkeys anoint themselves with the scent of an onion or a panda rolling around in Tabasco sauce. Further, this can open up dialogue with visitors about the thresholds of sensory systems in zoo animals.

Fig. 6.5 In nature orangutans thrive in an arboreal habitat (T. Maple)

Given that people rely so heavily on visual cues, it is not surprising that visual forms of enrichment are commonly used with great apes in the form of television shows or movies (see Clay 2011). It has been found that videotapes can be enriching for chimpanzees, although the effectiveness depends on several factors such as social housing condition, video content, and repeated exposure to videos (Bloomsmith et al. 1990; Bloomsmith and Lambeth 2000). In fact, visual enrichment can be used with a diverse array of species. Zoos utilize an improvised barrier called a "Howdy door" that provides visual access between animals who do not have physical access to one another. This involves aspects of social enrichment, and potentially auditory and olfactory stimuli, but the visual information may be important to some animals. While visual enrichment has frequently been deployed by zoo managers, it has received relatively little attention in research. In a study of laboratory chimpanzees, Parr (2001) discovered that her subjects responded to video representations of emotional scenes indicating they were capable of empathizing with onscreen peers as if they had actually witnessed it. Other studies have demonstrated that chimpanzees obtain information from video images and use this information to solve problems. Ogura (2012) provided video enrichment in the form of conspecific, human, and animated images to Japanese monkeys in a laboratory setting. Video enrichment was effective in reducing stereotyped behavior. When the subjects could control access and content they were highly responsive. Hoy and colleagues (2010) surveyed zoological institutions and found that visual enrichment was rated as the least important and the least used form of enrichment. This is a surprising finding giving the evident success of videography as an enrichment device. Hopefully future researchers will begin to address the potential role of visual enrichment in primates and other visually oriented taxa.

A unique example of complete sensory enrichment was practiced in the 1990s at the Jakarta Zoo in Indonesia. Zoo keepers routinely arrived at the exhibit with a horse-driven cart and picked up a group of orangutans for a supervised ride around the zoo.

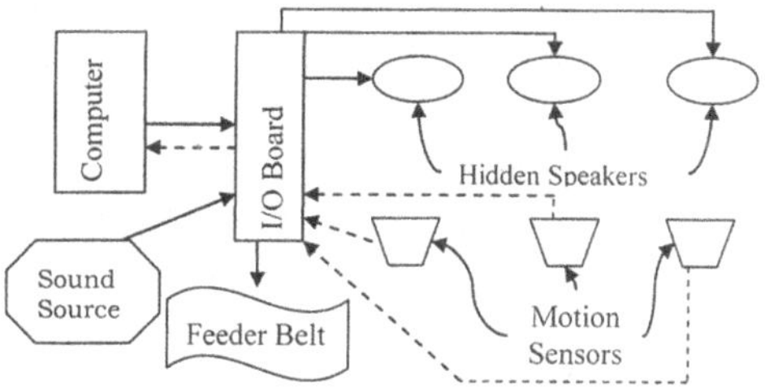

Fig. 6.6 Animal is rewarded for chasing the auditory stimulus (from Markowitz 2011)

The animals were provided with motion, auditory, olfactory, visual, and social stimulation by this opportunity to see the zoo and the people who were there to see them. This example conforms to the notion that visitors can provide a source of variability and stimulus change for animals and therefore qualifies as a form of environmental enrichment (Fig. 6.7). If this practice was objectively evaluated, we would want to know more about how the animals were managed during their brief public exposure. Whenever dangerous animals are put into close contact with people, the animals can be subjected to aversive handling techniques that generate considerable anxiety and fear. Some Asian zoos have utilized orangutans in photo sessions and advertised opportunities to join them for tea. It would be difficult to label these practices as good welfare. In China there are venues where giant pandas are also offered for close-up photos with visitors. Pandas and orangutans utilized in this way have been known to injure visitors, sometimes seriously. Handling dangerous bears, great cats, lions, leopards, and tigers, has resulted in many keeper deaths in non-accredited theme parks, and vicious attacks on visitors who strayed behind the scenes unsupervised. Regardless of the motivations of good-hearted caretakers, the organization that takes such risks is not practicing rational animal welfare (Fig. 6.7).

Basic raptor enrichment is a subject recently described on the website of the International Association of Avian Trainers and Educators (IAATE). Avian experts suggested that to accommodate their superior vision, captive raptors should have access to a perch with a view, especially a view of the sky. They also like to manipulate objects with their powerful feet and beak, while access to water provides them with an opportunity to bathe which all birds seem to enjoy. Many captive birds including raptors like to tear things apart so the presentation of cardboard boxes or tree cuttings provides stimulation and occupies time. Bird rehabilitators continuously exchange information on enrichment so their birds are busy even when they are tethered. Tethering limits the distance a bird can travel from its station, but enables the bird to preen, eat, bathe, extend it wings, flap, etc. IAATE supports the tethering of hawks, eagles, owls, falcons, etc. but does not recommend tethering non-raptors. In a zoo, however, raptors are not permitted to capture prey as hunting falcons do, but many are engaged in controlled flight during bird shows. Bird trainers know that flight is enriching and inactivity represents compromised welfare, so they must work with the birds throughout the day, even carrying them about the zoo on their arms to give them visual enrichment. IAATE recommends monitoring of all tethered raptors throughout the day to ensure health and safety. One problem with tethered birds, even injured birds in retirement or rehabilitation, is that visitors may perceive tethered animals as abused and neglected, but it should be noted that wild raptors

Fig. 6.7 A group of orangutans experience an enriching tour of the Jakarta Zoo (T. Maple)

spend most of their time perching not flying. Acceptable welfare for zoo raptors should include both opportunities to fly and safe settings for perching, so a zoo must have flexible surroundings for exercise and constraint. Unlike raptors, wild condors soar high in the air to locate carrion on the ground. In zoos they are generally presented in large flight cages that do not permit soaring but provide an acceptable quality of space. All vultures must be provided enrichment to keep them occupied. Condors and vultures benefit from whole carcass feeding. Those that are trained should be flown at least once each day (Fig. 6.7).

6.1.5 Social Enrichment

Given the many logistical constraints in place at zoos and aquariums, enrichment of the social environment may be one of the more challenging to implement, but important nonetheless. A social life does convey health benefits as field workers have determined in studies of baboons. Joan Silk (Silk et al. 2010) studied Chacma baboons (*Papio ursinus*) in Botswana, where she focused her research on 44 females. Female baboons having the strongest, most stable, and longest-lasting social relationships lived significantly longer than animals with fragile and unpredictable relationships. Silk and her colleagues suggested that friendship buffered the effects of stress and boosted physiological repair mechanisms conducive to longevity and good health. Complex social groups are clearly enriching for all group-living, nonhuman primates. Social enrichment in the zoo might involve the addition and/or removal of members of one's social group, if appropriate. However, this can sometimes introduce unnecessary aggression or may not be possible if animals are separated for reasons such as the prevention of breeding.

However, there are alternative ways to provide social enrichment that don't necessarily involve physical contact. As previously mentioned, "howdy doors," which provide visual, olfactory and/or auditory contact between individuals, may be used to socially enrich animals that must otherwise live apart. Other alternatives such as mirrors or videos of other animals may also help to broaden the social context. One of the great advantages of holding a large group of animals is the ability to form contiguous enclosures where conspecifics are a source of auditory, olfactory, and visual stimulation. The enriching nature of the four contiguous social groups of gorillas established at Zoo Atlanta in 1988 resulted in information that helped managers reconstitute social groups. The lone gorilla Willie B., an animal that had not yet been socialized, was clearly attractive to nearby females and they made their interest known by repeatedly orienting to him. Once socialized, Willie B. proved to be a highly effective breeder and parent. Of course, one type of social contact that almost all zoo animals encounter at least periodically is with humans, which brings us to the next enrichment category.

6.1.6 Human–Animal Interaction

Human–animal interaction is most commonly reported as interactions between care staff and animals. This can involve a wide range of interactions. For example, in a given day an animal will most likely interact with a keeper during feeding times and will be shifted between enclosures and moved for cleaning. Keepers and veterinary staff also play a role with training for basic behaviors and medical procedures. Positive reinforcement training techniques have vastly improved these interventions as zoo animals are now routinely trained to willingly present body parts for inspection. This training is also incredibly useful for the administration of medicine and other basic health manipulations without chemical immobilization. Another type of interaction that is becoming more common at the zoo is the animal-researcher interaction. This type may range from explicit "hands-on" training, to participating in a research project, to more indirect procedures such as behavioral observation. Even for observational research, the researcher is a part of the environment and may provide stimulation to the animal even if there is no direct contact.

In a multi-institutional survey of zoo keeper behavior from 46 participating zoos in North America, Carlstead (2009) identified two independent dimensions that describe animal relationships with their keepers; affinity to keeper, and fear of people. She noted that relationships with keepers are more favorable if the keeper interacts with the animal through barriers rather than entry into the enclosure. This finding speaks volumes about the value of protected contact in the management of zoo elephants. Furthermore, the ability to visually monitor the keeper is associated with affinity even if the keeper goes into the enclosure. Keeper attitudes can also influence their relationship to the animals they tend, suggesting that this will be an important area of future research.

Estep and Dewsbury (1996) cautioned that the presence of humans near the enclosures of breeding animals can interfere with normal sexual behavior. The effect

Fig. 6.8 Some insects easily and quickly adapt to handling (T. Maple)

is particularly troublesome if an animal develops fear in the presence of keepers, due to rough handling or some other developmental event. Because most captive animals have daily contact with keepers and visitors, the behavior of humans can have subtle but far-reaching consequences on sexual behavior and ultimately reproductive success. Whenever an animal fails to breed, zoo managers immediately suspect the influence of some traumatic past event in the animal's rearing history, particularly if it was not reared by its mother in a normal social group, e.g. the lowland gorilla Ivan who spent his formative years in a human household. For animals that must be hand-reared, the responsibility of the caretakers cannot be overemphasized. A growing literature describes the techniques that lead to reasonably normal socialization in a diversity of species (e.g. Read and Meier 1996; Bolwig et al. 1965; Bond and Block 1982; Hardin 1976; Kirchshofer et al. 1967; Maple 1980a, b; Wortman and LarRue 1974) (Fig. 6.8).

Szokalski et al. (2012) delivered a questionnaire to 86 zoo keepers worldwide to evaluate various methods of working with great cats. Protected contact was found to be the most commonly used handling method. Protected contact was also rated as more beneficial than hands-on or hands-off techniques for the cats, keepers, and visitors. Keeper respondents expressed perceived benefits including bonding between keepers and cats, and better educational outcomes. In addition to the obvious danger of hands-on contact with great cats, keepers reported that it delivered the wrong message to visitors. When compared to training, none of the handling techniques was rated as beneficial for all parties as training. Similarly, a questionnaire administered by Hosey and Melfi (2012) asked 130 zoo professionals to rate the strength of their relationship with zoo animals. The instrument was constructed to discover how bonds with zoo animals affected the animals and the caretakers. The survey revealed that human-animal bonds were widespread among keepers, scientists, and others working in a zoo, while respondents believed the bonds promoted good welfare for the animals and satisfaction for the caretaker. The association of good welfare and job satisfaction was thought to be a useful benefit of human-animal bonding in zoos.

Finally, another type of interaction between animals and humans at the zoo occurs when animals are on exhibit. Zoo visitors may influence an animal in an enriching or potentially distressing manner. Zoo and aquarium staff should monitor animal behavior in the context of any of these human–animal interactions in an effort to ensure that they remain enriching and not stressful. Most of the research on this topic indicates that visitors can be a source of stress for zoo animals (Fernandez et al. 2008). For example, Mallapur and Chellam (2002) observed that Indian leopards (*Panthera pardus*) were less active when visitors were present when compared to days when the zoo was closed. Leopard pacing increased on very busy days. In a classic study by Chamove et al. (1988), a variety of primate taxa were studied in the presence of visitors. Cotton-top tamarins (*Saguinus Oedipus*), Diana monkeys (*Cercopithecus Diana*), and ring-tailed lemurs (*L. catta*) exhibited increased aggression and a decrease in grooming, inactivity, and affiliative behavior when visitors were present. In a study of four groups of lowland gorillas, Stoinski et al. (2011) reported that crowd size had no uniform effect on activity or social behavior. There was some indication, however, that adult males may be more sensitive to crowd effect, while individual differences in personality may also influence reactivity in this species.

Gareth Davey (2007), a psychologist at the University of Chester (UK) recently published a comprehensive review of the effects of visitors on zoo animal behavior. One of the earliest indicators of visitor effects was reported by Thompson (1976) who observed increases in the sexual behavior of chimpanzees in response to higher numbers of zoo visitors. However, additional research has proved to be inconclusive in terms of the cause and effect of animal and visitor behavior. For example, instead of visitor density affecting zoo animals, the animals may be responsible for the change in density. Margulis and her colleagues (2003) at Brookfield Zoo in Chicago found no effect for visitors on zoo felids, but visitors were drawn to exhibits where the cats were active. Exhibits can be designed to facilitate visitor effects, the most common feature being the glass-fronted enclosure that brings the visitor closer to the animals (Fig. 6.9). In apes, the close visual inspection can generate animal-visitor play or aggression, from a safe distance of course. When researchers installed barriers to obscure the public view of a group of gorillas, aggression and stereotyped behavior declined (Blaney and Wells 2004). As Davey noted, there are flaws in much of the research on visitor effects; small sample sizes, a distinct nonhuman primate subject bias, and few manipulated variables. Of course, it is difficult to conduct experimental studies in the public environment of a zoo. Because it is intuitively obvious that visitors can have a negative effect on zoo animal behavior, curators, keepers, and veterinarians need to keep a careful watch for abusive forms of visitor behavior. Some zoos (Fig. 6.10) have designed creative signage to address this problem, prompting visitors to be quieter and more respectful. Open exhibits have to be constantly patrolled to retrieve objects that may be thrown to the animals. Training can provide additional protection as animals can be shaped to trade objects thrown into their exhibit for a food reward.

Fig. 6.9 Family encounter with polar bear at the Toledo Zoo (A. Norman)

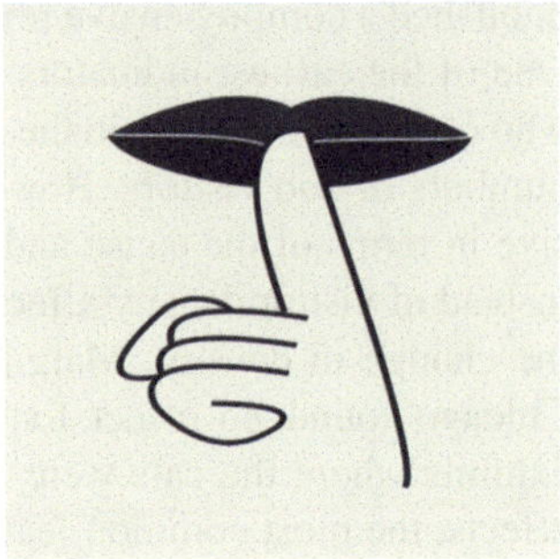

Fig. 6.10 Visual prompt to protect gorillas from excessive noise (San Francisco Zoo)

6.1.7 Cognitive Enrichment

Cognitive enrichment refers to the process of challenging and stimulating an organism's memory, decision-making, judgment, perception, attention, problem solving, executive functioning, learning and species-specific abilities. The types of cognitive enrichment range from simple manipulations to more complex, computerized test systems and can be implemented with almost any species. For example, an animal could be presented with a problem-solving task such as a foraging opportunity that promoted species-typical behaviors (e.g., extraction, manipulation, hunting). Computerized or manual tasks can be introduced to explore issues such as self-control, decision-making, social learning, memory, and perception.

This approach not only provides the researcher with invaluable information, but simultaneously enriches the animal by creating a stimulating and challenging environment in which it works for food and reinforcement. One study reported that chimpanzees engaged in a cognitive research program were more similar to wild chimpanzees in terms of activity budgets than captive chimpanzees not participating in cognitive research (Yamanashi and Hayashi 2011).

In the wild, animals with a high degree of cognitive capacity, such as chimpanzees, make choices about food items that are more than just nutritional decisions. According to Masi et al. (2012), chimpanzees in Kibale National Park in Uganda, consumed many foods with medicinal properties, including anti-tumor, anti-malarial, anti-bacterial, anti-diarrheal, and de-worming features. Apparently, wild chimpanzees learn to self-medicate from experienced members of their social group. Given their propensity to make these food choices, caretakers, nutritionists, and veterinarians might be able to structure a zoo diet that provides some of the medicinal properties they obtain in nature.

A study by McGowan et al. (2010) demonstrated that grizzly bears (*Ursus arctos horribilis*) were more attracted to concealed food that could be extracted from destructible objects than similar objects lacking food. Thus, foraging enrichment is a better intervention than the use of toys that lack functional relevance, according to this research. The bears also exhibited "contrafreeloading" propensities, choosing resources that require effort in the presence of freely available food. Although contrafreeloading (aka working for food) may be an artifact of captivity, it may also represent an adaptive form of exploratory behavior in seeking information about alternative food resources. We believe that creative contrafreeloading opportunities in zoos can be arranged to provide animals with choices and challenges in their immediate environment, contributing to their psychological well-being. A formal program in contrafreeloading at the Phoenix Zoo has been very successful, resulting in a management practice that requires work for food. Since animals vary in their capability to work, the zoo reached a working definition as follows: contrafreeloading is defined as diet offered with the help of devices, substrates, locations, and preparation to decrease or completely eliminate free feeding and extend foraging time. At many zoos throughout the world, great cats are activated by the introduction of bags or boxes scented with the urine of other animals. These items can also be propelled with ropes or pulleys to encourage predatory behaviors and stimulate cognitive processes.

6.2 Issues Regarding Enrichment

As with other enrichment classification systems (e.g., Bloomsmith et al. 1991), some types of enrichment could be classified as more than one category, but the important goal should be to attempt to incorporate all of these types at some point in an enrichment plan and provide caretakers with a framework from which to generate new ideas about enrichment. In doing so, one can create a complex, changing environment, one in which an animal has greater control and choice over its surroundings.

One of the most challenging questions to address is whether or not enrichment practices successfully meet the goals for which they were designed. In some cases, an enrichment protocol may be designed to address a very specific problem, such as a bear that has recently started to pace when on exhibit. The reduction or absence of stereotypical behavior is a common goal of enrichment programs. For example, in the case of the pacing individual, additional enrichment might be added to provide stimulation and alleviate boredom. If that was in fact the impetus for the pacing, enrichment may effectively reduce the undesired behavior and would be reflected in a significant decrease in the behavior. In fact, several meta-analyses, which compile the results across many studies, have found that enrichment effectively decreases stereotypic behavior. Swaisgood and Shepherdson reviewed enrichment publications from several journals and found that 53 % of the studies reported a reduction in stereotypic behavior. Shyne (2006) used meta-analytic techniques to analyze the effects from 54 studies and found a significant effect of enrichment compared to baseline conditions, with 90 % of studies exhibiting this expected pattern. Watters (2009) recently suggested that our inability to predict enrichment outcomes was due to the lack of a unifying theory. He observed that since animals evolved to cope with uncertainty in nature, the best path to successful enrichment will lead to uncertain and unpredictable rewards. Since animal activated delivery is cost-prohibitive, Watters suggested the use of timed delivery systems. To test the model, one-half of the animal's daily rations can be delivered in one place at a consistent location while the other half can be delivered randomly. Uncertainty in the context of enrichment should be more effective than the provision of enrichment items on a fixed schedule. Systematic research in this domain should lead to an integrative theory of environmental enrichment.

However, it is often the case that situations are not so straightforward. Returning to the example of a bear pacing, this behavior may have begun as a result of increased crowd size, the onset of breeding season, ambient noise such as construction, or any combination of these and other factors. In this case, enrichment may not influence the behavior given that it does not necessarily alleviate the underlying concern driving the stereotypy. For example, stereotypies may develop from situations in which animals experience unavoidable stress or fear, barren environments, or the lack of opportunities to fulfill species-appropriate activities. Furthermore, stereotypies may not reflect current conditions, but remain from past experiences. Further, animals exhibiting stereotypic behavior may actually be better at coping with environmental stressors than those exhibiting no stereotypies (Swaisgood and Shepherdson 2005). Nonetheless, there appears to be a relationship between stereotypic behavior and enrichment, and generally speaking, efforts to increase enrichment and alleviate these behaviors will benefit animal welfare. Careful attention to individual differences and systematic experimentation will be necessary to address these various concerns and systematically identify factors underlying stereotypic behavior, as well as other concerns such as an excessive reliance on enrichment without adequate attention to other factors.

One risk involved in enrichment programs is that if they are not carefully planned, implemented, and evaluated periodically, managers may rely too heavily on the "status quo" and become complacent. Although there is a great benefit to creating "logs" or calendars that prescribe when and what enrichment items should be given to a particular animal, if one only abides by this, without accounting for the animal's response, the point of the enrichment may be nullified. Simply tossing a boomer ball or Kong toy into an exhibit may count as enrichment, but we are obligated to determine if it is actually effectively enriching. Providing functionally useless toys or stimuli to animals is often a shortcoming of enrichment programs (Newberry 1995). Many zookeepers are dedicated to creating novel, exciting enrichment ideas that are tailored specifically to the animals they manage. What is most important is that these individuals are given the appropriate time and resources to conduct evaluations in order to avoid the risk of complacent enrichment.

This issue also arises in relation to exhibit design. Enrichment should not be considered a situational quick fix for poorly designed, understimulating environments. Ideally, exhibit design will take into account species-specific needs at the onset of the planning phase, and exhibits will be designed to be enriching. However, as is often the case, exhibits may be past their prime or delegated to species that they weren't originally designed for. In these cases, enrichment should not be used to preclude or delay renovations and alterations to the exhibit that would facilitate better animal welfare. In the worst case scenario in which alterations cannot be made, enrichment should be provided to the animals within the exhibit, but it is critically important to avoid an over-reliance on enrichment as an excuse to avoid necessary renovations and upgrades.

When feeding enrichment is provided to engage the visitor's paternalistic propensities, zoo managers must be careful not to over-stimulate the animals being fed. For example, many zoos are now offering visitors the opportunity to hand-feed giraffe, usually from a high platform that puts the visitor at eye level with the animals. This type of feeding has influenced group size as too few giraffe leads to a high level of stress. When the herd is large enough to enable rotation through feeding stations, the animals are subjected to an appropriate rather than an excessive amount of contact with visitors. Large giraffe herds contribute to welfare in several ways so this is a positive trend that also benefits visitors.

Regarding exhibit aesthetics, it is comforting to know that when enrichment is provided in a naturalistic environment, it doesn't necessarily detract from its appearance. Kutska (2009) found that un-natural items introduced as enrichment into a naturalistic polar bear exhibit, equipped with a 90,000 gallon water feature, did not adversely alter visitor perceptions of the enclosure, the zoo, or its conservation mission. Gorillas at the Basel Zoo in Switzerland were provisioned with shredded newspaper in such quantities that it seemed unnatural. However, given the hard features of this indoor exhibit the substrate became functionally soft and flexible (Maple and Hoff 1982).

Writing online in ZooLex, an information resource for zoo managers, zoo biologists, and zoo designers (Chap. 8), Fiby and Berthier (2007) suggested the goal of all planning for zoo animals should be to provide a maximum of activity and behavioral choices throughout the site. For example, in describing the master-plan of the Paris Zoo, the irregular form of the Aviary allows more complex and longer flight patterns and encourages investigatory flights by the birds. The varied landscape was designed to provide better opportunities for hiding food and other forms of enrichment. This exhibit has clearly been designed to improve the lives of the occupants.

The Paris Zoo plan called for extensive new plantings and protection of mature trees to take advantage of the many enriching features of living foliage. New water features were also planned to stimulate locomotion, exploration, and play. Mixed species exhibits take advantage of zonal separation, for example when terrestrial gorillas are housed with arboreal Colobus monkeys (*Colobus guereza*) and Sitatunga antelope (*Tragelaphus spekii*). The enriching features of exhibit rotation have also been developed for implementation at the Paris Zoo. The authors envision the use of rotation in the contiguous exhibits of lions and baboons as a way to introduce strong olfactory stimuli. As they argued, exhibit planning for enrichment depends on the full participation of key animal staff, curators, keepers, and veterinarians. No design team should program an exhibit without early input and continuous quality review by primary caretakers and other users of the facility.

Another enrichment strategy works particularly well for animals that are normally active at night; elephants, rhinos, hippos, hyenas, lions, tigers, etc. The best example of this idea is the "Night Safari" opened in 1994 in Singapore adjacent to the Singapore Zoo. Night Safari is an open-air zoo situated in a humid tropical forest on 40 ha (0.4 km^2) that is only open at night. The animals are lit by simulated moonlight so it does not disturb the normal behavior of crepuscular or nocturnal species. Because the ownership built Night Safari around an entertainment and dining plan, the business support elevated welfare for animals that prefer to be active at night. However, a plan such as this one could and should start with the idea of providing greater welfare with night opportunities provided for select species, much as the Oakland Zoo has achieved with elephants. To make night access affordable, a night restaurant or festive safari program can be built around the new visitor opportunity of seeing animals at night. In most zoos, a night safari program would involve a smaller subset of animals, or specific animals acquired for night exhibition such as African bushbabies (e.g. *Galago senegalensis*). A night exhibition program significantly increases the activity of animals that would normally be confined to night quarters. Many zoos now offer night safari programs, usually during the very hot months when night visitation is so compelling. For animals with particularly idiosyncratic habits, for example, the Asian fishing cat, Night Safari visitors can observe predatory behavior that is rarely seen during the day. When the first two fishing cats were born at the U.S. National Zoo in 2012, the mother and her kittens regularly entered the water to play and explore. Night access to a fish-provisioned pool produces an array of natural behavior in fishing cats.

6.3 Implementation and Evaluation

A final consideration is how enrichment practices are actually implemented. Hoy and colleagues (2010) conducted a thorough multi-institutional survey of enrichment use in captive mammals. As discussed in Chap. 3, multi-institutional studies provide a good approach to overcoming issues of small sample size and the authors have interesting data to report as a result. They surveyed 25 zoological institutions that utilized enrichment with mammals. Feeding enrichment was considered the most important, as well as the most frequently used. They also reported that the biggest deterrent to implementing enrichment protocols related to time pressures. Similarly, Clay and colleagues (2007) found that animal care staff considered limited time to be one of the constraining factors on implementing technological enrichment for captive animals. Moving forward, it appears that one of the challenges in establishing effective enrichment programs is to ensure that carestaff have the time and resources necessary to fulfill enrichment goals. Further, some form of monitoring should be encouraged, whether through behavioral observation or less formal daily logs. By gathering this information, one can be sure that effective enrichment techniques are identified and expanded, while obsolete or potentially stressful practices are noted and avoided.

In their partnership with the Wolong giant panda breeding center in Sichuan, China, zoo biologists from the San Diego Zoo and Chinese panda experts collaborated on a specialized enrichment plan. Enrichment was seen as a facilitator for reproductive behavior in animals that failed to breed, and a technique to modify an array of highly repetitive, idiosyncratic stereotypies such as pacing, pirouetting, head-tossing, self-biting, somersaulting, masturbating, swaying, tongue-flicking, sitting up, paw-sucking, cage-climbing, and regurgitating. The team worked on many approaches to increase social behavior and reduce abnormal behavior. They designed and built larger and more naturalistic enclosures; increased the structural complexity of existing enclosures; improved crowd control; encouraged more positive animal–keeper interactions; provided more bamboo and high-fiber biscuits that increased processing time; increased the frequency and variability of feedings; provided opportunities for interacting socially and chemically with peers; and provided manipulable objects to encourage the animals to work for food. These interventions largely worked, increasing activity and variability of social behavior, and reducing stereotypies. The enrichment program was based on the principle that animals need to exert control over their environment. In a discussion about their unique panda enrichment philosophy, Swaisgood et al. (2003) supported the concept of functionalism over naturalism, advancing the idea that choice and contingency may be more important than a strict adherence to naturalism. The San Diego panda enrichment program is comprehensive in scope and represents a useful template for enrichment in other species (Hare et al. 2003).

Research on enrichment at Brookfield Zoo (Whitham and Wielebnowski 2009) led to the development of a low cost, animal-based monitoring tool that enables keepers to quantify qualitative assessments of individual well-being. The staff is now able to utilize 12 species-specific Welfare Score Sheets in the monitoring process. The following species were included in the Brookfield program: Aardvark, African elephant, black rhino, clouded leopard, fennec fox, Goeldi's monkey, green-winged macaw, leopard gecko, okapi, polar bear, red-tailed hawk, and western lowland gorilla. As the score sheet system is tested and validated, it is expected that the instrument will be exportable to other institutions and encourage collaborative studies that expand the size of the sample.

Like welfare as a whole, enrichment studies have been skewed to favor studies of mammals, especially nonhuman primates. One taxonomic group that has been ignored is actually one of the best studied and most abundant of all animals; reptiles and amphibians. Over 6,000 species of reptiles and more than 4,000 amphibian species are currently recognized by scientific authorities. The most authoritative review of enrichment and welfare of amphibians and reptiles appeared in a book chapter by Hayes et al. (1998). Although we know a lot about many species currently living in zoos, their husbandry and management has been uncreative. Exhibits for these diverse taxa have assumed they need very little space and prefer a sedentary life. For the giant snakes and large varanids, in particular, such highly active animals require major improvements to reach a level of acceptable welfare in zoos and aquariums. Welfare scientists and zoo designers have an opportunity to produce innovations that will significantly improve welfare given the complexity of amphibian/reptile sensory capabilities, and the urgent need for improvements. Unfortunately, the widespread belief that reptiles and amphibians have a narrower cognitive and behavioral repertoire and their significant genetic distance from mammalian forms renders them low priorities for environmental enrichment.

The most creative thinking about reptile enrichment and training has been generated by Gordon Burghardt, a professor of psychology at the University of Tennessee. Burghardt's website (utk.edu/~gburghar/) demonstrates the depth and breadth of his ideas in the many types of husbandry he has suggested for reptiles. According to Burghardt and his collaborators (1996), reptiles are capable of play and they benefit from problem-solving challenges presented by their keepers. Sensory enrichment is also good for reptiles especially when the scent of prey items are introduced into tubes and other objects placed into their enclosure. The larger the species the more challenging it is to provide enrichment. The largest lizards are the monitors (*Varanus* spp.) and they are considered to be the most intelligent. Burghardt provided a link to a the Colchester Zoo, UK (www.colchester-zoo.co.uk) which has constructed a remarkable exhibit for Komodo dragons (*V. komodoensis*). According to a description accessed in September 2012, the animals in the "Dragons of Komodo" exhibit are able to utilize a wide variety of environmental innovations including a selection of soft substrates to encourage digging, a massive glass roof that can be opened on sunny days, and a system that provides rain showers simulating weather conditions in Indonesia. At the National Zoo in Washington, D.C. a variety of food is regularly provided as an enrichment strategy for many of the reptiles. For example, keepers have reported that Komodo dragons can take up to 3 h to consume an entire rabbit, occupying time as it would in predatory encounters in the wild. Similarly, the ubiquitous boomer ball has been introduced to Aldabra tortoises (*Aldabrachelys gigantea*) with some success, while all tortoises can be shaped to investigate a food puzzle.

Giant snakes are generally inactive in zoos but their huge size makes them a favorite of zoo visitors. Sensory enrichment introduced into elevated translucent tubing would provide a kind of rodent "habit trail" (Trademark of the Hagen Corporation) for constrictors. A large habit trail for an adult reticulated python (*Python reticulatus*) or anaconda (*Eunectes murinus*) would provide a tunnel to investigate and hopefully activate the snake while educating and entertaining zoo visitors. Of course, the innovation would have to take into account their nocturnal habits, and may work better at night as a special demonstration for education programs or special events. Giant snakes could be activated in nocturnal houses if provided motivating stimuli. Like rodents, snakes occupy warrens and tunnels and are therefore a good candidate for this type of habitat. Reptiles would be far more interesting if zoo designers would introduce activity as a design criterion for these creatures. On the internet the reader can find video records of anacondas in the wild. In one video produced by National Geographic (Anaconda hunt) the snake is filmed as it hunts a young capybara in the water. An exhibit for giant snakes might offer verticality so the animal could climb as it would in the wild. Reticulated pythons moving up a tall tree would be enriching and exciting to observe. Anacondas would benefit from a water feature that would encourage exploration and reveal their great size and swimming skills. Reticulated pythons and anacondas are the largest snakes, both species reaching 30 ft or more in length, while the green anaconda has reached a weight of more than 500 lb. Over the years, zoo herpetologists have put together very creative micro-habitats for smaller reptiles that looked natural; the next frontier is to create habitat that influences behavior in reptiles of all sizes and shapes.

In a landscape immersion setting, giant snakes will be most compelling suspended in trees where visitors will have to visually search for them. Designers will need to plan for the full spectrum of the snake's behavioral repertoire and not just for the moments when it is quietly digesting its meal. Similarly, in large aquatic environments, it is possible to give sea turtles an opportunity to swim fast over a diverse terrain. What is missing is the opportunity for a sea turtle to engage in active foraging and feeding so typical of its life in the ocean. A designer working with enrichment in mind, and a staff prepared to train them, could develop a very entertaining operant system for these highly intelligent and curious animals. The inspiration for our thoughts on this issue is a continuing field study of Hawksbill turtles (*eretmochelys imbricata*) in the waters of South Florida. Larry Wood's videography demonstrates that this species is constantly searching for its favored food items, using its strong beak to tear vegetation from crevices in the substrate. It is enriching for turtles to feed this way in captivity and a powerful educational device. Naturalistic exhibits in zoos and aquariums have enormous potential to meet the psychological needs of species that eagerly enter water (Figs. 6.11 and 6.12).

Invertebrates have also received attention from keepers intent on improving their quality of life in the zoo. At the Smithsonian National Zoo, where enrichment is a priority of every animal unit, new objects are frequently introduced to the curious giant octopus; hermit crabs are frequently offered new shells for habitation; new foods are given to crayfish; spiny lobsters get unopened clams that represent a challenge; and a carved pumpkin is offered as entertainment to a colony of roaches at Halloween.

Mellen and MacPhee (2001) made a number of salient recommendations about enrichment programs. First, enrichment must be proactive based on the animal's natural and individual history, and exhibit constraints. By optimizing the environment, abnormal behavior can be prevented. Second, there must be a plan to generate natural behavior. If the institution cannot provide adequate enrichment, it may be necessary to

Fig. 6.11 Anaconda welfare is improved by water features (Shutterstock.com)

Fig. 6.12 Water features activate Malayan tigers (K. Lovett)

eliminate this species from the collection. Third, evaluations must be objective based on standard data collection techniques. Fourth, there must be accountability for consistent planning, implementation, execution, and documentation of enrichment, and ultimate responsibility resides with the institutional CEO. Directors must insist that the programs are conducted properly if enrichment is going to succeed.

In this chapter we have highlighted some of the critical factors influencing the design, implementation, and assessment of environmental enrichment. In the coming years, this subject will be of continued relevance to welfare and interface with innovative exhibit designs that also activate the residents. We anticipate that researchers, managers, and planners will continue to dedicate human and financial resources to the synergistic relationship between zoo design and animal husbandry.

Chapter 7
Behavior Analysis and Training

The problem of occupational therapy, of providing pastimes in the zoo . . . has recently been given more and more attention, to the great benefit of the animals. We are no longer content to let animals vegetate stupidly in narrow cages, as in the old-fashioned menageries, but do our utmost to see that their lives are healthy and full, and as positive as possible.
H. Hediger

One key specialty that needs to be unified under the zoo animal welfare banner is "behavioral management." Going by different labels, including behavioral husbandry and applied behavior analysis, behavioral management is a discipline of psychology that deals with the functional analysis of environment and behavior. It is also comprised of the training technology associated with the science and practice of operant conditioning. Behavioral management is an umbrella term that encompasses many aspects of welfare, including behavior analysis, environmental and cognitive enrichment (Fig. 7.1), environmental design, biological (ecological) constraints, socialization and re-socialization techniques, and both medical and performance training. Zoos that once depended on former circus and carnival trainers discovered better practices and better training standards in the realm of science. The circus tradition utilized aversive control techniques and featured trainers who physically dominated and sometimes abused lions, tigers, bears, and elephants. Some of them, in the circus tradition, fired starter pistols and snapped whips to get the attention of the animals in their show. Fortunately, traditional circus training methods have been largely discredited and they are no longer practiced in modern, accredited zoos. Marine mammal training, in aquariums, marine parks, and some zoos, is an entirely different tradition, based on positive control and the use of both food and tactile reinforcement. Trainers in aquatic shows typically interact with their subjects in a playful and more respectful manner. However, marine mammal training has also been criticized for subordinating the animals, and for its use of demeaning story lines. Circus training, with few exceptions, never produced any research to further our understanding of mammal behavior and cognition, although Hediger studied circus

T.L. Maple and B.M. Perdue, *Zoo Animal Welfare*, Animal Welfare,
DOI 10.1007/978-3-642-35955-2_7, © Springer-Verlag Berlin Heidelberg 2013

Fig. 7.1 Chimpanzee operating a simple "termite probe" to obtain mustard (T. Maple)

animals for insight into the psychology of captivity (Hediger 1955). However, aquariums once employed or collaborated with behavioral scientists whose work (often supported by federal research grants) significantly advanced our knowledge of dolphins, whales, and other marine mammals (e.g., Kellogg and Rice 1966; Tavolga 1966; Pryor et al. 1969; Herman 1987). Hediger took a special interest in the marine mammal training program at a pioneering Marineland park that he visited in Florida in the 1950s. This facility, located in St. Augustine, Florida, is now owned and operated by the Georgia Aquarium. In his book, *Studies of the Psychology and Behaviour of Animals in Zoos and Circuses*, Hediger noted the connection between training and play in the trained dolphins at Marineland:

> There are imperceptible gradations between playing and training . . . these two things are not opposites. Good training is disciplined play. Both play and training often give excellent opportunities for brightening up the daily existence of the animals in the zoo, making it more significant, and giving the animals the necessary amount of exercise and occupation. (p. 139)

Unfortunately, training programs in zoos and aquariums have become disconnected from their scientific history. It is generally the case that zoo and aquarium trainers are the experts who we deploy to train others, and rarely are the trainers fully vetted in the science and practice of operant conditioning. Some of these expert trainers have vast experience and cannot be faulted for their superior skills at training any zoo or aquarium species. However, we believe that the training culture that exists today lacks sufficient depth to reach its full potential. Certainly the utility of training to advance welfare requires a complete understanding of all of the tools available in the behavior modifiers toolkit. It is time for zoos and aquariums to reconnect with the academic wellspring of a virtual universe of promising welfare strategies.

Given the need to strengthen the scientific foundation of training, the Association of Behavior Analysis International (ABAI) is one organization of psychologists that could become a valuable resource for zoo and aquarium professionals. While a number of zoo-oriented symposia have been organized for ABAI conferences, so far the zoo profession hasn't invited many ABAI experts into our midst for a constructive dialogue. Because academic psychology departments are no longer populated by a critical mass of operant conditioners, zoo biologists must search for specialized collaborators who have the ability to keep our training programs dynamic and relevant. Years ago, nearly every psychology department employed at least one faculty member committed to the operant paradigm of B.F. Skinner, the dominant research approach in psychology. But behaviorism is not dead; it is still an active if hidden specialty in psychology, largely because operant conditioning works, as Roediger (2006) recently observed:

> We know how to alleviate or eliminate phobias through extinction ... we can reduce problematic behaviors and increase the probability of desired behaviors by judiciously providing and withholding reinforcements ... behavioristic analyses exist in self-management programs, in industry, in sports, in parenting guides, and of course in animal training programs for pets and for zoos. Anywhere that prediction and control of overt behavior is critical, one finds behavioristic analyses at work. (p. 23)

In a relevant contribution by Bloomsmith and her associates (2007), human behavior modification techniques were evaluated for their efficacy as a tool in treating nonhuman primate behavioral problems. As they concluded, virtually all of the techniques found to be effective in treating human stereotypy and self-injurious behavior are directly applicable to the same behaviors in nonhuman primates. With this in mind, behavior analysts working in human clinical settings can be recruited to work side by side with zoo keepers, zoo biologists, and zoo veterinarians. In many cases, abnormal behavior in humans, monkeys, and apes share a common etiology. Likely as not, behavior therapies developed for use with humans will be effective for many different species. Of course, the best collaborators are behavior analysts that have animal experience, but those who lack animal training are generally keen observers and capable of finding the right solutions in an active collaboration with knowledgeable zoo professionals.

7.1 Mentors and Partners

A sophisticated knowledge of training and its potential as an enrichment and stress-reducing tool is an extremely important background for zoo curators, veterinarians, and keepers. Unless we find a way to recruit and hire for this specific training, national associations such as AZA may have to employ consulting behavior analysts to offer formal training to our curators and key zoo staff. As this book goes to press, zoo and aquarium caretakers have been invited to participate in a spring 2013 course on "Animal Training Applications in Zoo and Aquarium settings" at Disney's Animal Kingdom. The lecturers are all zoo biologists with extensive knowledge and experience in training exotic animals, but this course and future courses can be strengthened by seeding the teaching team with professional behavior analysts from academia or private practice. In the meantime, we would like to see a series of formal workshops and symposia that feature behavior analysts and zoo biologists addressing animal welfare issues together. Partnerships of this kind would likely produce new approaches to lingering problems. Operant conditioners are needed in the zoo and aquarium field, as collaborators, consultants, or in dedicated staff positions. As Forthman and Ogden (1992) observed, the functional relationship between environment and behavior is the forte of applied behavior analysis:

> This kind of analysis results when knowledge of ethology is combined with knowledge of conditioning to identify and quantify the environmental stimuli functioning in operant and respondent conditioning. (p. 648)

This combination is illustrated by a study carried out by Altmann et al. (2005) who switched zoo lions from the conventional 6-days a week feeding schedule to a "gorge and fast" schedule that simulated feeding patterns in the wild. The lions were slowly introduced to feeding 3 days each week but received the same amount of food as before. Not only did the animals exhibit an increase in appetitive (goal-oriented) behavior, but pacing was cut in half. Digestibility was also improved, so there were nutritional gains as well as behavioral consequences. In a study of three species of bears that received frozen "fishcicle" enrichment, Forthman et al. (1992) also achieved a reduction in repetitive pacing.

Behavior analysts working on stereotypies in human subjects have determined that it is always difficult to completely eliminate stereotyped behavior, but operant techniques can significantly reduce their frequency and duration. When confronted with a very persistent pacing problem in a zoo animal, a certified behavior analyst would be qualified to devise management tactics to modify the behavior. We would do this in the same spirit of inquiry that would prompt a zoo veterinarian to recruit a consulting physician from a hospital known for its expertise in cardiology. Fortunately, there are some consulting firms that are already working with zoos by providing behavior analysis and behavior modification services. One such firm, "Active Environments" has an excellent track record of solving difficult animal management problems in zoos and aquatic parks throughout the world. They have tutored animals to cooperate for medical procedures, avoid hot-wires, and adapt to change; intervened to modify abnormal behaviors and aggression; and shaped isolates and socially-deprived animals to interact with peers. They are also the primary developers of the protected-contact system of elephant management, providing the training techniques for working with elephants without entering their enclosure (e.g. Desmond and Laule 1991). In our opinion, protected-contact has been one of the most significant advances in zoo management in the past 30 years, although the innovation initially met fierce resistance within the zoo profession. Regarding the free contact alternative to their model, Whittaker and Laule (2009) observed:

> ...keepers would never routinely enter the same space with a bear or rhino, yet it is done with elephants. It would be considered unacceptable for a keeper to strive for social dominance over a snow leopard, yet free contact keepers attempt to establish and maintain this type of relationship over elephants. It would be unacceptable to routinely employ the techniques of negative reinforcement and physical punishment with gorillas, yet these techniques are used regularly with elephants in a free contact system. (p. 181)

Free contact elephant managers may disagree with this characterization, but given these contradictions, it is not surprising that most zoos have converted to elephant management protocols associated with the protected-contact model. Furthermore, the Association for Zoos and Aquariums recently advised member institutions that it is unsafe to occupy the same space with elephants, although zoos reserve the right to enter in an emergency. Many experts believe that protected contact is also better and safer for the elephants. The efficacy of operant conditioning is demonstrated by the fact that trainers who have moved from free to protected contact have experienced effective compliance by elephants now managed by distant voice control (A. Stone, personal communication).

In a study designed to compare elephants exposed to both free and protected contact systems of management, Wilson and Maple (unpublished) found that a small group of African elephants adapting to protected contact initially complied less frequently with verbal commands delivered by their keepers. It was apparent that these elephants were beginning to exercise limited control over their environment, after years of compliance in a free-contact system, and these findings suggest the emergence of "autonomy." Because free contact promotes a culture of dominance where the trainer exerts control over the elephants, the distancing of the trainers and the concomitant removal of aversive control is a contribution to improved welfare. In a recent publication, we offered a strong argument against the use of aversive control techniques with elephants (Maple et al. 2009) noting that aggression is the most troublesome side-effect of aversive control. Our position agrees with the position statement on the website of the Animal Behavior Management Alliance (ABMA), an organization of animal care professionals using behavior management to advance the welfare of animals:

> The Animal Behavior Management Alliance (ABMA), in support of our core values, does not endorse the use of aversives in routine animal management. Physical or psychological intimidation increases fear, hinders learning, can increase aggression, and is detrimental to animal welfare.

Gail Laule, the co-founder of Active Environments, deploys behavioral management as a combination of positive reinforcement training and environmental enrichment techniques. Laule and her colleagues have developed new protocols for modifying abnormal behavior patterns in a variety of zoo animals (www.activeenvironments.org). Her work always begins with observation. The behavior in question has to be carefully delineated; when does it occur? Is there a discernible pattern? What variables seem to influence it? Once enough data are gathered, she formulates a working hypothesis. In the case of a throwing or spitting animal, is the behavior related to situational stressors, or is the animal seeking attention? In a case like this one, the conventional wisdom is to extinguish the bad behavior by failing to reinforce it, while providing salient reinforcement for a competing, acceptable alternative, e.g. vocalizing for attention. To sustain the good results obtained with training, both trainers and keepers must continuously monitor their subjects (Fig. 7.2). In our work with Active Environments in Atlanta, we knew that we would have to be vigilant after the consultants completed their assignment. Training should be consistently refreshed by experts, and there is value in keeping them or someone like them involved as consultants to the management team (Fig. 7.3). One of the firm's most impressive

Fig. 7.2 Hippo responds to a target at Disney's Animal Kingdom (Walt Disney Co.)

examples of a behavioral management strategy resulted in the socialization of a sub-adult male drill (*Mandrillus leucophaeus*). The Active Environments investigators determined that the source of stress in this drill group was a moderate state of social and sensory deprivation resulting in the expression of abnormal behaviors and a low level of pro-social behaviors. By systematically increasing sensory stimulation through formal training sessions, providing reinforcement by successive approximations for eating and relaxing in close proximity to others, and by establishing feeding stations and targets where keepers could interact with the animals, the group demonstrated significant increases in all forms of positive social interactions, and abnormal behaviors were reduced (Laule 1993).

A functional analysis procedure was also used by Martin et al. (2011) to reduce the frequency of aggressive throwing and spitting by a chimpanzee. Caretakers did not respond to the undesirable behavior, but rewarded a competing, socially desirable behavior, in this case vocalizing.

There are other interest groups that are organized around the discipline of training including IMATA (International Marine Animal Training Association) and IAATE (International Association of Avian Trainers and Educators). These groups have also embraced zoo animal welfare and environmental enrichment. On IAATE's website many enrichment tips have been offered (e.g. Chap. 6). The association is particularly concerned about disseminating current knowledge, standards and practices. In all of these groups, there is a need to connect those who care for zoo and aquarium animals and those outside experts who understand the scientific principles of behavior analysis.

Historically, aquatic parks with marine mammals were among the first to utilize operant conditioning. This was a great step forward because Skinnerian positive control techniques were highly effective with dolphins, sea lions, and birds. Oceanariums in California, Florida, and Hawaii long ago pioneered the use of operant conditioning to administer medications and to draw blood from marine

Fig. 7.3 Zoo keeper inspecting elephant's foot, demonstrating compliance achieved through the management system of protected contact and positive control techniques (G. Laule)

mammals without anesthesia. One of the first zoos to utilize positive control was the U.S. National Zoo when they hired marine mammal trainer Karen Pryor (www.clickertraining.com) to work with keepers to train great apes and other species. Pryor not only trained animals, she also trained keepers, and she did so in part to inspire them. Given the drudgery associated with the keeper position, the National Zoo administration wanted to motivate their staff to take an intellectual interest in working with exotic animals. Training represented a new set of skills and an opportunity to improve the quality of life for many species. Animals were trained to come closer to the visitors, clean up their enclosures, weigh themselves on a scale, and enter or leave enclosures on voice command. As Pryor (1981) concluded:

> Positive reinforcement training constitutes an exchange of deeds for goods in which a pleasant communion arises, a salutary sort of equality between animal and trainer. One cannot work without the other, and both must do their part. That is reinforcing in itself, for both parties ... operant conditioning in zoos is not only a benign addition to animal management practices, but rewards staff and animals alike. I am sure that B.F. Skinner would not be surprised to hear that, he would just wonder why it took so long. (p. 98)

Looking to the history of psychology for insight, Tarou and Bashaw (2007) provided tested principles from applications of the experimental analysis of behavior (EAB), a Skinnerian system of research findings, to guide enrichment and training programs in the zoo. They recommended extrinsic reinforcement techniques (provision of food, social access, etc.) to increase desired behaviors. They noted further that if keepers make extrinsic rewards more difficult to obtain or provide more or higher quality reinforcement, it will likely increase the long-term success of enrichment. They cautioned that care staff should avoid continuously reinforcing behavior after the response is established, enrichment immediately after feeding, or exposing animals to enrichment when reinforcement is no longer available. EAB is a vast literature containing valuable procedures and techniques that can be utilized in shaping the behavior of zoo and aquarium animals. Tarou and Bashaw concluded that the utility of EAB principles requires further validation through an active program of zoo research on contingencies and schedules of reinforcement.

7.2 Scholars and Leaders

A program of behavioral management and research is especially important for institutions that operate educational shows, for example, aquariums and marine parks with dolphins and whales, or zoos that train performing birds. In our opinion, the future of interaction with exotic animals, and this concern applies to elephants, depends on a healthy and dynamic culture of scientific training. We are so concerned about the disappearance of animal learning labs and operant conditioning courses in college and university curriculums that we have suggested major investments in endowed chairs and centers devoted to the history and science of training. Training through operant conditioning (positive control) contributes to animal welfare by providing an intellectual challenge and generating physical activity, so public training sessions marketed as theater can be justified as long as they are not demeaning or too demanding. Objective behavior analysts could participate in the internal evaluation of captive facilities and programs, and stimulate innovations that enhance welfare. A trainer employed by a sea park cannot be completely objective, but an endowed professor in an independent or associated think tank must be. Endowments and partnerships with universities and specialized academic centers and institutes will significantly upgrade our collective intellectual capital. Aquariums are serious about training, but the current generation of trainers is not grounded in research, and few of our trainers have advanced degrees in psychology. Compared to the founding era of marine theme parks, in the 1950s and 1960s, science flies below the radar screen in modern aquariums, but an active scientific program specializing in behavior analysis and training could provide the expertise to significantly advance animal welfare.

Aquatic parks that exhibit dolphins and whales provide thrills for millions of visitors each year, but they also receive criticism from both animal welfare and animal rights groups. A recent survey by the Humane Society of the United States (2012a), the Animal Welfare Institute, and the Whale and Dolphin Conservation Society has shown for the first time that a majority of Americans are uncomfortable with killer whales (orcas) in captivity. This is not surprising given the fact that these highly active marine mammals live in very small enclosures when compared to the vast stretches of ocean where they evolved in the wild. The poll found that the inability of orcas to engage in natural behaviors and the negative consequences of confinement in small pools are sufficient reasons to stop keeping orcas in captivity, according to the respondents. Although the surveyors are hardly objective about zoos and aquariums, it is difficult to argue with the assertion that killer whales are just too big, too intelligent, and too demanding for the provision of acceptable welfare in zoological facilities as they are currently configured (see also Rose 2004). As many zoos have dramatically improved their facilities for elephants, aquariums will need to literally "think outside the box" to provide for the basic psychological needs of large whales. For those highly intelligent and social marine mammals currently living in captivity, frequent interaction with trainers provides essential cognitive and social stimulation. A resumption of meaningful cognitive research in these facilities would provide a higher level of enrichment for the animals when they are not on stage and significantly improve their welfare. A commitment to a program of cognitive research would also elevate our conversations about the efficacy of specific welfare action. The *Journal of Applied Behavior Analysis* (*JABA*), recently published an article that called for more research papers that demonstrate the utility of behavior analysis in solving animal problems (Edwards and Poling 2011). For highly intelligent marine mammals, it can be argued that behavior analysis is the solution to boredom. We anticipate renewed interest in this field if zoo biologists and behavioral scientists engage in meaningful scientific partnerships focused on the science of aquatic animal welfare.

Combining the applied orientation of experienced aquarium trainers with the scholarship of an academic unit of certified behavior analysts would likely generate new ideas and help to solve persistent problems. Even the best zoos have not investigated all of the possibilities to strengthen their expertise and creativity in animal care. The daily demand of operating our public institutions slows husbandry innovation to a snail's pace. We need academic and medical partnerships to hasten change. Connecting every zoo training unit to a psychology department in a nearby university is a feasible first step but the mutual benefits of such partnerships have to be carefully delineated. Partnerships are always strongest when these connections are made at the highest possible levels; deans, provosts, or presidents. The advantages to universities are direct access to exciting research settings with opportunities to study charismatic mega-fauna, the potential for collaborative funding from private sources and foundations, and the opportunity to publish and present original research findings. Vastly underestimating their value to the university, zoos and aquariums tend to think too small about the possibilities, and ask for too little in return for the provision of their talent and expertise and access to their unique collections. Behavior analysts who partner with zoos will find fertile ground for their research, exciting work opportunities for their students, and the reward of making a substantial contribution to the conservation and welfare of exotic species. An additional benefit is the likelihood that collaborations of this kind will breathe new life into behavior analysis and stimulate innovation in zoos and aquariums.

There are academic outposts of behavior analysis throughout the world. In the United States, some colleges and universities still maintain a strong specialization in the experimental analysis of behavior and applied behavior analysis, e.g. Western

Michigan University, University of the Pacific (Stockton, California), and the University of North Texas. At UNT, the Organization for Reinforcement Contingencies with Animals (ORCA) was founded in 1999 as a unit of the Behavior Analysis specialty in Psychology. A student organization, ORCA is very active in organizing conferences and workshops. Behavior analysts are found in small colleges and large universities, often engaged in entrepreneurial relationships with institutions that house exotic animals, e.g. Rebecca Singer, a professor of psychology at Georgetown College in Kentucky, and Eduardo Fernandez, a post-doctoral scientist at the University of Washington and Research Fellow at the Woodland Park Zoo. A wider utilization of these and other experts in behavior analysis and training should be encouraged.

7.3 Implementing Behavioral Management

Bloomsmith (1995) described a sophisticated behavioral management program at the M.D. Anderson Cancer Center housing large colonies of chimpanzees and rhesus monkeys supported by the U.S. National Institutes of Health. The Texas facility instituted one of the first formal enrichment programs administered by scientists and managed by a dedicated "environmental enrichment technician." Bloomsmith also deployed a fulltime trainer and hired consultants to help develop a formal training regime, and organized the entire program into four distinct teams for enrichment, training, research, and cross-training. The latter made sure that all parts were interchangeable. One of the most intriguing ideas that emerged from this team-approach was the installation of a mister-system that the animals could operate on their own. Water as enrichment has been tried before, originating with the work of Hal Markowitz who designed showers under the control of Asian elephants at the Portland Zoo. Markowitz' original idea has spread to benefit other creatures in the zoo. For example, the Louisville Zoo recently opened their Gorilla Forest exhibit including space devoted to pygmy hippos (*Choeropsis liberiensis*), equipped with a shower that the hippos can operate themselves. The great success of the Texas primate facility is instructive and the lessons learned there can be applied to zoos. As Dr. Bloomsmith concluded:

> Over the last eight years, we have elevated what once was a part-time activity for some to something that is a full-time occupation for many. I have no doubt that the greatest beneficiaries of this increased effort are our rhesus monkeys and chimpanzees. They are now the recipients of a better-organized system, with more knowledgeable people delivering more varied and more challenging enrichment . . . For these reasons; I think our primates are living in less stressful and more stimulating environments. (p. 10)

A recent publication (Tresz 2006) reviewed the behavioral management program at the Phoenix Zoo, a useful example of what can be accomplished in a zoo setting with applications of behavioral technology. The program was originally implemented to achieve proactive standards for the basic care and psychological well-being of the living collection, and to meet or exceed the guidelines of the U.S. Animal Welfare Act. The behavioral management goals at this zoo were established to:

> ...encourage nonhuman animals in the collection to use their natural abilities, to promote species-appropriate behavioral and mental activities, and to offer a sense of self by allowing choice and experiential exploration. (p. 65)

Tresz documented the commitment of zoo keepers to protocols that derived from an internal animal behavior and behavioral management philosophy. The program seeks to involve staff at all levels of the organization. Indeed, zoo staff utilized behavioral techniques to emphasize stimulating, interactive exhibits rather than those that just looked naturalistic. Environmental changes included the addition of a deep layer of sand for elephants, and a new wallow, scratching post, and sandstone boulders. Elevated hay and browse increased foraging time. The elephants were activated by commands to repeatedly walk from one end of the exhibit to the other, vigorous exercise that was reinforced by the keepers. A comparable training procedure was used with alligators at the Palm Beach Zoo (see Chap. 4) as a strategy for weight loss and improved wellness. Organizational changes at the Phoenix Zoo facilitated innovations in behavioral management. A behavioral management coordinator position was created to provide leadership and accountability for the program. Volunteers were recruited to provide sufficient human resources. The zoo also hired consultants with expertise in training, exhibit design, and behavior to assist with the planning and implementation of specific projects. In addition, the program is now monitored and supported by the involvement of a formal behavioral enrichment committee. At the Phoenix Zoo, behavioral management is a key strategy in animal care and clearly a factor in the improvement of zoo animal welfare, but their academic partnerships with nearby universities have waned in recent years. The nearest university program specializing in behavior analysis is located at Northern Arizona University. Faculty at NAU could be recruited to help teach operant techniques to zoo staff, or take on specific problems familiar to certified behavior analysts, e.g. repetitive locomotion, hyper-aggression, etc. A useful way to approach collaborative problem-solving would be a weekly or monthly seminar where problems could be introduced and debated, and solutions can be proposed and implemented by faculty-guided graduate students and zoo staff. Nearly every zoo is located close enough to a community behavior analysis unit to set up exchanges that contribute to more effective and professional behavioral management programs. We can only hope for a revival of behavior analytic interest in zoos and aquariums, but zoo professionals must issue the invitation for this to happen on a formal and continuing basis.

Some zoos have enjoyed extraordinary success in the application of operant training techniques. One example is the operant conditioning training program at the Bronx Zoo for New World primates (Savastano et al. 2003). This work unit focused on seventeen species (88 callitrichids and small-bodied cebids) in 26 social groups. The animals received training for hand-feeding, syringe-feeding, targeting, scale and crate entry, and transponder readings. Zoo staff recorded habituation to husbandry procedures, improved compliance, and a diminution of aggressive threats toward care staff. They concluded that the stress of intervention had been reduced and welfare enhanced. When visitors were present the increase in voluntary

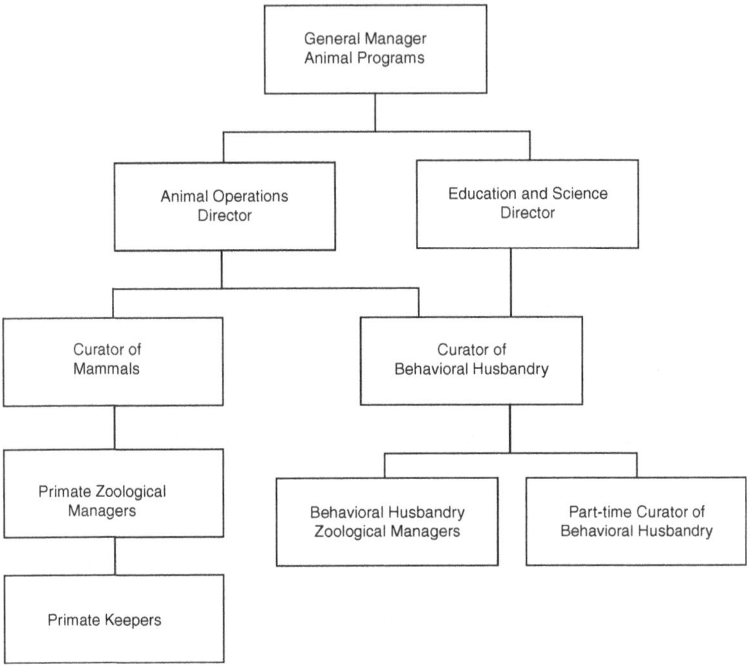

Fig. 7.4 Reporting relationships at Disney's Animal Kingdom (Mellen and McPhee 2001)

interaction with keepers enabled education about the practice of low-stress husbandry and the value of operant training. The authors' concluding remarks are suitably optimistic:

> ...despite the various challenges involved in the development of a formal training program for a large, diverse primate collection, success can be attained if the goals of the program are prioritized and the available resources maximized. More important, if enhancing the welfare of the animals remains the primary objective of the program, then the challenges encountered become the stimulus for new solutions. (p. 259)

Pomerantz and Terkel (2009) examined the effect of positive reinforcement training (PRT) on the well-being of 12 zoo-housed chimpanzees. The investigators recorded a significant decrease in abnormal and stress-related behaviors and a significant increase in prosocial affiliative behaviors following implementation of PRT. Further, lower ranking individuals benefitted to a greater degree than high ranking subjects. Pomerantz and Terkel suggested that PRT could be considered a type of enrichment that produced lasting general effects even when the training had nothing to do with behaviors indicative of welfare. In other words, positive reinforcement training contributes to welfare because it is inherently enriching.

At Disney's Animal Kingdom (DAK) the staff implemented an operant training program early in their history. Their reliance on training is extensive, based on empirical guidelines designed by Mellen and McPhee (2001). The organizational structure at DAK (Fig. 7.4) provides the expertise and leadership that we recommend for all zoos that utilize operant conditioning and behavior analysis techniques. DAK recruited doctoral level scientists and administrators to build their state-of-the-art animal management structure. The Disney team continues to innovate and publish extensively.

When giant pandas (*Ailuropoda melanoleuca*) arrived at Zoo Atlanta in 1999, Bloomsmith and her collaborators (2003) utilized positive reinforcement training to encourage pandas to move around their enclosures. Training was effective in generating movement as it has been for other animals, and the training process generalized to all times of day producing consistent, reliable control over movement. Giant pandas that learned to cooperate with commands enabled staff to keep their enclosures clean, provide fresh bamboo, conduct non-invasive veterinary inspections, or engage in cognitive testing, thus benefiting the psychological well-being of the animals. Panda trainers invested approximately 45 min per week for each animal for 7 months of training, and saved an enormous amount of time in moving pandas throughout their enclosures during daily management and husbandry sessions. Once trained, the animals immediately transferred their compliance to new trainers. The experience of other zoos in San Diego and Hong Kong demonstrated that training could be achieved in other giant pandas with little difficulty. Surprisingly, their specialized food habits did not detract from their enthusiasm to comply with trainers. Bamboo worked well as a reinforcer and so did the presentation of leaf-eater biscuits.

Giant pandas are highly intelligent animals as Tarou et al. (2004) demonstrated in a study of giant panda spatial learning. Tarou's investigation was a pioneering evaluation of a species that had not been previously studied to determine its capacity to learn. Because of their charismatic personality and natural beauty, opportunities to observe research with pandas should be a popular venue with visitors. At such moments, keeper talks can focus on conservation, education, and animal welfare, and the steps zookeepers and veterinarians are taking to keep the animals healthy and well. Pandas too are responsive to training that renders them compliant for a wellness examination that educates and informs the public. With this species zoo educators do not have to worry about getting and holding the attention of zoo visitors. Figure 7.5 illustrates the use of shaping to train giant pandas to open their mouths on command, a behavior that facilitates health monitoring by keepers.

7.4 Behavior and Welfare at the Oakland Zoo

With a strong institutional commitment to applied behavioral research, Oakland Zoo curators and keepers systematically altered their elephant exhibit and their management protocols to improve elephant welfare (Kinzley 2009). Specifically, the team wanted to increase foraging and feeding time, daily distance traveled by elephants, and reduce the incidence of abnormal stereotypic behavior. To accomplish their goals, Oakland management decided to provide elephants access to their yards at night during the 8 months of mild weather in Northern California. Overall physical activity and species-typical behaviors increased as a result of this change

Fig. 7.5 Shaping giant panda Yang-Yang to open mouth at Zoo Atlanta (A. Thompson)

of protocol. Further, stereotypic behavior was nearly eliminated in the hours the elephants spent outdoors at night. Because elephants were provided food at night, feeding and foraging time was increased by 3.5 h. The total feeding/foraging time at the Oakland Zoo was increased to nearly 12 h, much closer to the norm of 12–18 h that wild elephants spend feeding and foraging in a 24 h period. Remarkably, enrichment interventions including dispersing browse increased distance traveled in the enclosure by fourfold, from one-half mile to an average of 2 miles per day. An additional feature to the Oakland elephant management program was the expansion of outdoor space from 1.25 to 6 acres, providing the elephants with greater control over their exposure to shade, visitors, and habitat variation. The Oakland Zoo, like Atlanta, was once regarded as one of America's worst zoos; it is now widely respected for putting the animals first. Oakland Zoo's innovations in elephant care and commitment to superior standards and practices has boosted its reputation as a national leader in zoo animal welfare. Their enhanced reputation was not achieved with money alone, but with innovation, creativity, hard work and the will to sustain meaningful change.

7.5 Behavior Analysts in the Zoo Workplace

Enrichment is an interesting blend of field biology and operant conditioning and it is regarded as a valuable tactic in strategic behavioral management. Veterinarians who regularly interact with zoo animals benefit from the training that accompanies enrichment. Animals willingly present their arms or legs to give blood, hold themselves still for close visual inspections, examinations by stethoscope or ultrasound, or submit to recording other instrumented biological data. A veterinarian's job is

Fig. 7.6 Tractability of black rhino enables operant enrichment strategies (Zoo Atlanta)

much easier when animals comply with intervention and examination, while the animals suffer no risk from invasive anesthesia (Fig. 7.6).

Because zoo staff assigned to train animals may not fully understand the principles or theoretical foundation of behavior analysis, formal instruction on the psychology of operant conditioning is essential training for zoo biologists and veterinarians. During the senior author's tenure at Zoo Atlanta, zoo managers and affiliated graduate students collaborated with a national leader in behavior analysis, M. Jackson Marr, a professor in Georgia Tech's School of Psychology. Professor Marr taught advanced courses in behavior analysis and operant conditioning to all graduate students conducting research at Zoo Atlanta during the formative period of our research partnership. Over the years he also organized symposia for student collaborators to present papers at the annual meetings of the Association of Behavior Analysis International (ABAI) in Chicago, Phoenix, Orlando, and other cities. His training courses offered to undergraduates and supervised by graduate students at Georgia Tech utilized zoo animals in unique ways (e.g. Lukas et al. 1998). We hesitate to label Dr. Marr a "guru" of behavior analysis, but his key role as a mentor and collaborator was a comforting feature of the partnership between Zoo Atlanta and Georgia Tech.

A misunderstanding of classical behaviorism has led zoo administrators to label certain research personnel as "behaviorists." Our colleague and collaborator, Donald G. Lindburg, a primatologist who was a Professor of Anthropology before he was recruited to serve as the San Diego Zoo's first "Chief Behaviorist" is an example of this basic misunderstanding. Behaviorism is a specific school of psychology linked to the early research of Johns Hopkins psychologist J.B. Watson that became the dominant paradigm in psychology long before Harvard psychologist B.F. Skinner became its main academic proponent. A certified behaviorist in a zoo would be qualified to train animals though the application of operant conditioning techniques, and to study animals with the tools of behavior analysis. Behavioral management and behavioral husbandry are derived from operant psychology. Similarly, veterinarians often use the label "behaviorist" to describe any scientist who studies animal behavior. Although his previous specialty was primate behavior, Professor Lindburg studied cheetahs (*Acinonyx jubatus*) and giant pandas among other species at the zoo and at field sites in Africa, India and China. If we were in a position to hire him today, we would offer him the position of "Director of Animal Behavior," or perhaps "Director of Ethology." In this capacity he might supervise the work of a dedicated staff behavior analyst, or a consulting psychologist with training and expertise in the narrower specialty of applied behavior analysis/behavioral management. We aren't offended by the label "behaviorist" but it is more accurate to apply the label "ethologist" or "comparative psychologist" to describe the full scope of our interests and expertise. The reputation of the San Diego Zoo has been greatly enhanced by its association with eminent behavioral scientists. Duane M. Rumbaugh, one of the twentieth century's most important authorities in comparative cognition initiated his studies of great ape learning at the San Diego Zoo in 1954 while he was a professor of psychology at San Diego State College (Rumbaugh and Washburn 2003). A series of anthropologists, psychologists, and zoologists from San Diego State and, more recently, the University of California at San Diego have utilized the San Diego Zoo collection as a naturalistic laboratory for behavior research (e.g. Segal 1989).

An experienced field scientist, Don Lindburg's keen understanding of wild animals led him to arrange whole carcass feeding so felids at the San Diego Zoo could experience an enriching consumatory response. He also deployed mechanical, inedible lures that cheetahs could pursue and capture in a large enclosure designed to simulate hunting (Lindburg 1998). This was a highly creative innovation that has been widely imitated. Behavioral management (or behavioral husbandry) has become a key specialty in federal primate research centers, and zoos are increasingly consulting with specialists in psychology whenever inexplicable behavioral problems arise (e.g. Maple et al. 2009; Martin et al. 2011) or whenever operant techniques can benefit the animal's welfare. Workshops to teach training methods to zoo keepers, curators, and veterinarians have been occasionally hosted by zoos and aquariums. Dr. Mollie Bloomsmith (now at the Yerkes National Primate Research Center) and Gail Laule (Active Environments) have teamed up to teach these unique workshops at many venues. We strongly recommend that zoos and aquariums committed to operant training methods should utilize experienced training professionals (e.g. Bloomsmith-Laule, or Karen Pryor) in a workshop format to augment and sharpen the skills of keepers who train on a daily basis. Many zoos utilize the expertise of Steve Martin who has produced free flight programs for raptors that operate as a seasonal concession. His company (s.martin@naturalencounters.com) provides a connecting point for researchers who want to collaborate with one of the world's most experienced animal trainers. He too provides services for installing behavior management programs in

Fig. 7.7 Karen Pryor with clicker-trained wolf at Wolf Park, Indiana (M. Sloan)

zoos and aquariums, and advises on innovations in enrichment for a variety of species. Needless to say, Martin regards free flight programs as a form of enrichment.

In 1996, staff at the Whipsnade Zoo (UK) utilized operant training techniques to shape greater one-horned Asian rhinos (*Rhinoceros unicornis*) to accept veterinary foot-care. Two years later they trained black rhinos to tolerate blood collection and ultrasound examinations without anesthesia. Rhinos proved to be very compliant and cooperative in adapting to the training process. They worked for both food rewards and physical contact from keepers who interacted through the barred-barriers in their night houses. From the references cited and the description provided in the paper describing their work (Holden et al. 2006), the protocols followed those developed in American zoos. One reference cited the work of Karen Pryor (1999), a highly experienced trainer and operant psychologist. Pryor's iconic book (*Don't Shoot the Dog*!) should be read by every zoo keeper, curator, and veterinarian. Pryor's mentorship of trainers and keepers popularized the use of clickers to modify zoo animal behavior. She is also an example of the value of true expertise. On her website, she explains the difference in techniques such as "shaping" and "successive approximation" (www.clickertraining.com). This level of understanding can greatly improve zoo animal welfare by identifying the quickest route to success in training. No one has done more than Karen Pryor to enable caretakers to effectively utilize operant conditioning to advance zoo animal welfare (Fig. 7.7).

In discussions with trainers at aquarium facilities in North America, we discovered a clear consensus that aquariums and zoos recognize the benefits that accrue from greater exposure to the experimental analysis of behavior. Our interviews affirmed their interest in rebuilding these relationships with traditional academic mentors. While many zoo biologists who studied psychology in graduate school could teach operant conditioning principles to keepers, this is not the level of expertise that is needed to make real progress. Experienced behavior analysts have toiled for years to master the complex literature of operant conditioning. These are the professionals we should recruit to the cause to take full advantage of the science of behavior analysis. Professionals from the outside will only strengthen the expertise and services of our dedicated trainers and caretakers.

Melfi and Hosey (2011) have also concluded that good welfare can be achieved through the application of positive reinforcement training. However, as they discovered, many institutions lack the knowledge to implement behavioral husbandry, and this deficiency compromises welfare. To remedy this situation the authors organized a workshop designed to introduce behavioral husbandry as a means to improve animal welfare. Sixteen delegates from institutions in developing countries (e.g. Brazil, Cameroon, Congo, Kalimantan, Sumatra, etc.) were fully funded to attend the workshop held at the Edinburgh Zoo in 2008. It was duly noted that capacity building for improving zoo animal welfare lags behind wildlife conservation, demonstrating there is a need to reach out to zoos, sanctuaries, and managed parks to help them assimilate new management techniques. This workshop successfully introduced the concepts of animal welfare, environmental enrichment, and behavioral husbandry to staff in positions of responsibility with the opportunity to advance welfare in the sixteen countries represented. Partnerships with established conservation organizations might serve as a bridge to governments that need to be educated about the efficacy of animal welfare. Hilda Tresz, Curator of Applied Behavior at the Phoenix Zoo, has been successfully engaged in mentoring small zoos in Africa, Asia, and South America, utilizing the behavioral management philosophy developed in Phoenix. Her mentoring programs have been designed for ease of communication and translated into several languages. The need for tutoring in behavioral husbandry outstrips the resources for people like Hilda, so this is an opportunity that must be advanced through targeted philanthropy.

The field of behavioral analysis is expanding throughout the world. This growth is demonstrated by the membership of ABAI which has recently experienced a doubling of its membership outside the United States. The membership of ABAI is now distributed among 30 countries. Jack Marr (2006), one of the most active ABAI leaders offered the following observations on the state of behavior analysis worldwide and its potential to revitalize the movement:

> . . .there is no doubt that many academic programs in behavior analysis [in the United States] are struggling to maintain their integrity. But in the rest of the world, many are not contaminated by doctrinaire and ignorant prejudices about the value of behavior analysis – as a science and a practice: they see that it works, and they want more. Thus the future of behavior analysis may well depend on our efforts in international development. (p. 17)

Needless to say, the interest in behavior analysis around the world should lead to new opportunities to collaborate with zoos and aquariums. The current interest in funding scholarships and certification in behavioral analysis is one way to partner academic programs and zoos. The fresh insights of students and faculty specializing in behavior analysis and zoo animal welfare could make a real difference, particularly in the developing world where deficiencies of zoo and aquarium facilities are well documented. Synergistic relationships are feasible through the regional associations of ABAI, organized zoo associations such as SEAZA in Asia, and entrepreneurial institutions such as the Whitley Wildlife Conservation Trust in the UK. We regard the brokering of these relationships as a high priority for advocates of a global zoo animal welfare movement.

Throughout the world, the front line of training and behavior analysis is occupied by zoo and aquarium keepers who literally do the heavy lifting each and every day. Behavior modification is more of a science than an art, so keepers need the proper training to be able to apply behavior change techniques. We trust that communication and cooperation between behavior analysts and frontline caretakers will be encouraged by zoo administrators. In working with any species in the zoo, VPI professor E. Scott Geller's (1985) behavior analytical training model is worthy of emphasis and a fitting summary to this chapter:

(1) Define the target behavior; (2) Observe the target; (3) Record the rate of occurrence; (4) Intervene with a program that changes consequences; (5) Compare the frequency before and after intervention; (6) Evaluate the findings. By effectively implementing scientific training methods, animal keepers are significantly improving the quality of life for zoo and aquarium animals.

Chapter 8
Designing for Animal Welfare

They are not brethren, they are not underlings: they are other nations caught with ourselves in the net of life and time, fellow prisoners of the splendor and travail of the earth.
Henry Beston

Zoo design throughout the world continues to be inspired by the words of Henry Beston (1928) who so eloquently articulated the familiar mantra that engenders such deep respect for the natural world. Building on Beston's remarkable insight, Robert Sommer (1974) understood that animals and people alike are victimized or venerated by the quality of their communities. Sommer's dichotomy between hard and soft architecture delineated the deficiencies of many traditional institutional settings including mental hospitals, prisons and traditional zoos. In a recent publication (Sommer 2008) he extrapolated from an earlier analysis by psychiatrist Henri Ellenberger (1960) who reviewed the parallel history of the zoo and the mental hospital. In eighteenth century England, as Ellenberger noted, the nation's first mental hospital, Bethlem (known also as "Bedlam"), was as much a tourist attraction as the London Zoo. Some writers characterized Bethlem as a "human zoo" where "oddities and characters" were on display. Under public pressure to change, zoos and mental hospitals have continued to evolve, as Sommer observed:

> In the best cases, the zoo developed into the wild animal park with natural habitat, discreet display, and animals in natural groups . . . separated from the public by moats rather than bars . . . In the mental hospital field, new pharmacological treatments shifted the emphasis . . . from hospitalization to outpatient treatment. (p. 378)

The key similarity among mental hospitals, prisons, and traditional zoos are the powerful negative effects of confinement and sensory/social deprivation. Over time, a life in hard confinement induced unusual if not bizarre behavioral adaptations (classified as deprivation acts) such as catatonia, coprophagia, regurgitation/reingestion, stereotyped rocking and pacing, head-banging, and other self-injurious behaviors. These behavior patterns are frequently idiosyncractic in form; for example, polar bears and other swimming mammals are prone to developing swimming

T.L. Maple and B.M. Perdue, *Zoo Animal Welfare*, Animal Welfare,
DOI 10.1007/978-3-642-35955-2_8, © Springer-Verlag Berlin Heidelberg 2013

Fig. 8.1 Hard architecture at the zoo teaches the wrong ideas about wildlife (T. Maple)

stereotypies in aquariums and zoos. A severely disturbed patient in California's Stockton State Hospital exhibited stereotyped locomotion backwards. The stereotyped behavior of confined autistic and developmentally disabled humans are known to be a function of prolonged social deprivation, providing vestibular, tactile, visual, and auditory stimulation not provided by caretakers (Baumeister and Forehand 1973; Thelen 1981). Psychologists continue to study the similarities and the differences in the way animals and people respond to confinement in mental hospitals, prisons, and zoos. However, a revolutionary new approach to zoo design is transforming the zoo from a hard to a soft environment conducive to the development of normal, natural patterns of behavior. In a paper published in *Natural History*, Sommer (1972) argued that hard zoo environments were producing abnormal animal behavior that served to distort the character of wild animals. In the wild, monkeys and apes don't bite themselves, repetitively body-rock, or consume their own feces. Hard architecture conceals the animal's true nature (Fig. 8.1). In this sense, hard zoos fail as educational settings. The antithesis of the hard zoo is defined by the progressive design movement known as "landscape immersion," an approach pioneered and branded in the 1970s by the Seattle design firm of Jones & Jones (Hancocks 2001). The design document produced for the Woodland Park Zoo may have been the first of its kind to assert that the animals were the primary client. Hediger, of course, was way ahead of all of us, as he long ago advised that zoos be designed for the animals, the caretakers, and the visitors, in that order. The revolutionary plan launched in Seattle shook up the status quo. Indeed it was the first shot fired in a war between the old school and the new school of zoo design. The new school had its detractors, but the landscape immersion concept spread quickly and widely, helped along by new design firms led by former Jones & Jones colleagues. Architects collaborate with curators, keepers, educators, and scientists when they begin the journey of planning a new exhibit. A good starting point is the list of priorities "Five Features of Zoo Design" in Table 8.1. In feature

Table 8.1 Five features of zoo design (after Seidensticker and Doherty 1996)

1. Select species that are exhibitable in the facility available.
2. Ensure that the macroniche is considered in the design; be certain that it is environmentally appropriate.
3. Establish and explicitly state which behavioral system is to be featured while the animal is in the exhibit space. Optimize the opportunity for non-aggressive social interactions during public viewing hours.
4. Provide species-specific resting or refuging sites.
5. Manipulate food type, amount, distributing and timing of deliveries to optimize vigilance, food-seeking and feeding behavior.

number three, Seidensticker and Doherty recognize that animals are sometimes exhibited as a subset, e.g. a pair rather than the group. The key to this feature is exhibit flexibility. In feature five, care is taken to activate the animals by distributing food in natural ways, unpredictably and in hidden locations to encourage searching and foraging.

In a cogent discussion of zoo design at the sixth Paignton Conference, Veasey (2005) presented his "Inclusive Habitat Model" of zoo design where animal welfare considerations are paramount and based upon the science of animal welfare and a deep understanding of behavioral ecology. In this model, the design must cater to the needs of all stakeholders in the following descending order; (1) The physical and psychological well-being of the animals is the first priority; (2) Animal care staff (keepers and medical staff) must have the tools to sustain health, welfare, reproduction and conservation; (3) The needs and preferences of zoo visitors, scientists, and society in general must be considered when designing exhibits; (4) The mission of the zoo must be served and advanced by zoo exhibits and facilities. In building a zoo, Veasey, like Hediger before him, dares to put the animals first. This idea must be revisited and codified as zoos have tended in recent years to emphasize the visitor experience over the animal's needs (Melfi et al. 2005). Of course, animal needs and visitor experiences are synergistic and need not compete for priority in the zoo. One could argue that Carl Hagenbeck Jr.'s pioneering, nineteenth century "panorama" exhibits were designed to provide naturalistic space for charismatic mammals, although its effect on the visitor was equally spectacular. He was the first to place multispecies groups in an appropriately themed landscape, the forerunner of today's immersion exhibitory, an idea that enriches the experience of animals and visitors.

8.1 Field Biology and Zoo Design

One of the earliest and best examples of design ideas derived from field experience is the contribution of Adrian Kortlandt (1960). Among his recommendations was the idea that zoos should let chimpanzees themselves choose the temperature they prefer by giving them an open enclosure combined with an indoor room with access to both. From his observations of wild chimpanzees, Kortlandt suggested the construction of zoo environments comprised of large and undulating open enclosures rather than simply a cage with trees. Within the exhibits, he recommended multiple male and female groupings so that animals would behave normally. He wasn't troubled by the likelihood of aggression in complex social groupings as he recognized the buffering value of emotional displays:

> A pursued male may defecate in fear, but he always runs faster than his pursuer. In a zoo, such a show will certainly not be less impressive than a group of roaring lions, for instance. (p. 78)

From his extensive field experience, Kortlandt believed that outbreaks of aggression could be managed by providing escape routes throughout the enclosure. The site could not have dead-end corners and there must be opportunities to retreat and hide, e.g. protective bushes, shallow water, obstacles, or multiple branches extending into an elevated hide. These features remind us of Hediger's (e.g. Maple and Perkins 1995) observation that the worst feature of many zoos is the corner of a cubic enclosure. According to Hediger, the "cube" was a false starting point for zoo architecture, as it is completely unnatural. In a naturalistic setting, to facilitate aggressive emotional displays, Kortlandt envisioned the provision of swinging artificial trees on springs, large drums, extended chains that rattle, and other noise-making equipment. Kortlandt also recommended mental stimulation with interactive public programs but he recognized that the theatrical approach could be perceived as demeaning. As Kortlandt concluded: ". . . the well-being of chimpanzees in zoos requires something more than what suffices for creatures lower on the evolutionary scale . . . chimpanzees need also intellectual employment." (p. 80)

Another experienced field biologist shared his expertise with zoo biologists in a paper published in the *International Zoo Yearbook*. A.H. Harcourt (1987) offered constructive ideas based on his significant field work on mountain gorillas. His commentary on the typology of zoo exhibits for gorillas is candid:

> The fault lies partly with the zoo management, which sometimes appear to disregard knowledge gained from studies of species in the wild, and partly with the field worker, who disregards both the special problems of captivity and the necessity of making his knowledge available to the zoo manager. This article is an attempt by one field worker to improve the situation for one species . . . (pp. 248–249)

Harcourt described the forest environment of wild gorillas as a complex, three-dimensional landscape. In these forests gorillas are known to spend 45 % of their time feeding on over 50 plant species in a home range of 5 km or more. The vertical dimension is important in nature as apes spend considerable time feeding and playing above ground. Field workers generally agree that captive environments lack sufficient environmental complexity. Restricted space also limits an animal's ability to escape aggression so attention must be given to engineering opportunities for dispersal, as Kortlandt advised. Great ape exhibits around the world have benefitted from the advice of experts such as George Schaller, Jane Goodall, Dian Fossey, and other lesser known but highly competent advisors. David Hancocks

(2012) believes that good design requires complete objectivity from referees and active, constructive criticism:

> Appoint an animal representative to every design team – ideally from outside the zoo – and give them responsibility to ask hard questions. 'What can I do for mental stimulation in this space? Why are you giving me a concrete bedroom with no windows and fluorescent lights?

Although the potential of landscape immersion has not been fully realized, many contemporary great ape exhibits are highly creative, and continue to demonstrate a commitment to greater welfare for these complex and demanding creatures, a result of the growing collaboration between zoo and field biologists. For example, Scotland's Edinburgh Zoo, long an innovator in Europe, has developed a naturalistic chimpanzee exhibit branded as "Budongo Trail". They promote the exhibit as a "state-of-the-art chimpanzee facility that links the excitement of seeing chimps up-close with wildlife conservation, science, and education". The exhibit is operated by the Royal Zoological Society as a partnership with the Budongo Conservation Field Station in Uganda. Their indoor-outdoor habitat is designed to house up to forty chimpanzees in a naturalistic colony.

The exhibit also provides educational space for visitors and school groups. Budongo Trail demonstrates how welfare is advanced when exhibit features are planned as simulations/replications of natural systems. Scientists and graduate students at nearby universities provide contemporary information and generate new ideas based on research. Working with zoo staff and community volunteers they also participate in monitoring the efficacy of the habitats. In this way, designing for welfare also promotes conservation for the groups that have served as the model for the zoo exhibit, a good example of "social marketing". Zurich Zoo's Masoala Rainforest simulation is another example of how a zoo can support field conservation (Fiby 2005). One third of the funding necessary to operate the actual Masoala National Park in Madagascar is provided by Zurich Zoo donations. The 11,000 m^2 of the Masoala Rainforest exhibit at Zurich Zoo contains 470 ambassadors for the Madagascar ecosystem. Once again, we can see how conservation and animal welfare are synergistic, advances in one clearly benefitting the other.

8.2 Verticality in Zoo Design

As we have seen, monkeys and apes frequently move about in the vertical dimension of space to obtain food, escape predators and social adversaries, to play, and to make comfortable nests for sleeping. Some species, South American spider monkeys and Asian gibbons, for example, are specialized to hang and swing through the air with ease. The arm-over-arm locomotion characteristic of gibbons and siamangs is known as "brachiation." Among the great apes, the most vertically specialized species is the orangutan, the largest primate that lives in trees. Of all the apes, orangutans have suffered the greatest degree of environmental poverty in traditional zoos where their vertical propensities were largely ignored by early zoo designers. Field data gathered by scientists in the 1960s and 1970s led to enlightened zoo architecture that provided orangutans with opportunities to climb high in artificial structures. Zoo orangutans that grow up in treeless environments may become habitually oriented to the ground, so tactics that induce climbing may have to be implemented. In a zoo that elevates their orangutans, this large, reddish ape becomes a post-card symbol for creative landscape immersion (Illustrated in Fig 5.5). On the floor of a cage with no climbing structure, orangutans look depressed and demonstrate little activity, the ultimate zoo couch potato. Reacting to the state-of-the-art in a bygone era, the humorist Will Cuppy (1931) was moved to comment: "The psychology of the orangutan has been thoroughly described by scientists from their observations of the sea-urchin." Their propensity to relax is unique but apparently natural to the orangutan. In a recent study, Pontzer et al. (2011) discovered that orangutans living in a large indoor/outdoor habitat used less energy, relative to body mass, than nearly any eutherian mammal ever measured. Although the authors suggested the finding may indicate an evolutionary trend for this species, no data exist to see if wild orangutans are this inactive. We are inclined to think it may be an artifact of captivity. If orangutans are low-energy specialists it may require a different strategy to activate them. For this species, inactivity may not be a true indicator of poor welfare. From observations of captive orangutans for more than two decades, we have observed serious and vigorous pursuit when adult males show interest in females. Females can be activated by hormonal peaks, exhibiting aggressive sexual behavior (proceptivity), as Maple et al. documented in 1979. They may be wired to be stoic and passive but they are certainly capable of vertical gymnastics given the opportunity to climb and engage in social displays.

In our opinion, the ultimate zoo habitat for this species has not been achieved. While verticality is the dominant trend, a superior zoo habitat would resemble an Indonesian rainforest with orangutans at a distance high in trees or climbing structures. Field studies have revealed that large males often come to the ground, so attention must also be directed to a soft, manipulable substrate. Sadly, the disappearance of rainforest trees throughout its range, forces orangutans to come to the ground just to navigate their damaged ecosystem. Drawing from the experience of field stations in Borneo and Sumatra, developed to rehabilitate former captives, zoo orangutans could be rewarded for coming down from the trees for close-up viewing by paying customers. Under these conditions, a keeper talk could explain the dire conservation challenges that wild orangutans face, while demonstrating the keen intelligence and complexity of this great ape. A zoo habitat with these unique features would be inspiring to visitors while providing optimal welfare for orangutans of all ages. Distance and depth provides the opportunity to utilize or engineer tall trees or structures that function as trees, and the greater volume will enable a greater number of orangutans to share the space. An exhibit of territorial males connected to females through narrow traveling tubes that only the females could enter would provide the opportunity for females to select their sexual and social partners (see Maple and Finlay 1986, 1987). Such a design would also facilitate vocal displays characteristic of adult males. Welfare-oriented zoo directors should always think big, take a deep breath, and think bigger about an optimal vision of forest living primates in zoos.

Great welfare is only achieved by a willingness to take calculated design risks. The optimal orangutan exhibit would likely become one of the most successful exhibits in any zoo, revealing the full potential of this spectacular red ape.

Perth Zoo in Western Australia opened an innovative, functionally optimal exhibit for Sumatran orangutans in 2002. Perth is known for their success in breeding this species, but the new facilities provided greater attention to details that support animal welfare. The designers started with one enclosure where the furniture was tested by introducing just one orangutan to occupy the prototype. According to the Perth report (www.e-architect.co.uk, January 2009), indicators of animal welfare improved in the new facilities. Unique features included flexible ropes and poles in multiple habitats, water cannons that orangutans could direct at their peers, and foraging tubes loaded with jams and jellies that could be extracted with a "termiting" tool. The nesting platforms were 12 m high, providing the apes a great view of the entire zoo. Although the climbing structures have a technical look and feel, they function much like a rainforest environment. The steel trees and ropes (branches) can be adjusted by the keepers to suit the use of orangutans of differing sizes and strength.

8.3 Activity-Based Design

Jon Charles Coe is one of the leading lights in the architectural domain of "activity-based design". He was a key player in the landscape immersion epoch when he worked for Jones & Jones in Seattle. Later he and his partner Gary Lee teamed up with Zoo Atlanta scientists to develop a revolutionary gorilla exhibit, the Ford African Rain Forest, opened in 1988. The firm evolved to become CLR, based in Philadelphia, and Coe and Lee designed every habitat innovation at Zoo Atlanta from 1985–2002. In 1985, Coe contributed the first design paper ever published in the journal *Zoo Biology* (Coe 1985) providing illustrations of design mistakes that compromise welfare. For example, as Coe demonstrated, old-school exhibits surrounded the animals with onlookers. These concrete grottos, ubiquitous in many California zoos, provided visitors with a 360 degree perimeter with a great view for the visitors and a lifetime of stress for animals. Further, the visitors themselves perceive the animals as victims, particularly because large crowds surrounding animals are often loud and abusive. A better design gives viewers segmented viewing nodes interrupted by vegetation, much as you would see the animal in nature. In Fig. 8.2, we have reprinted Coe's original drawing that depicts two exhibition models, one with the lion positioned above the visitors, and the other with the visitors in the elevated position. When located below the visitors, the exhibit invites abusive behavior and an attitude of visitor superiority. According to Coe, a lion exhibited above visitors gives the animal the position of dominance, and likely evokes a feeling of awe and respect.

Both Jon Coe and Gary Lee have continued to think outside the box. In an article published in the *Journal of the American Veterinary Medicine Association*, Coe (2003) acknowledged that immersion exhibits alone do not guarantee fitness or

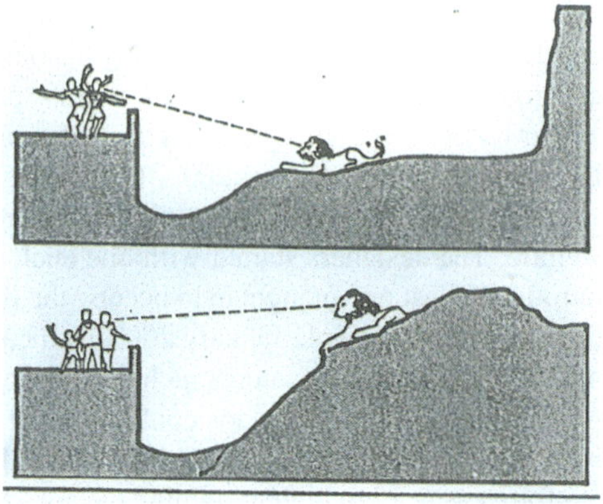

Fig. 8.2 Jon Coe's (1985) hypothesis that position affects visitor attitude

wellness. Many published studies have demonstrated that even social groups housed in immersion exhibits don't necessary utilize all the features intended to promote health and welfare. He offered a working definition of the concept of activity-based design:

> . . . {it} begins with the premise that the animals' long term well-being is paramount and that environments, programs and procedures which advance this goal are frequently of great interest to the visiting public. Healthy animals with stimulating behavioral choices tend to be more active. Therefore, opportunity-rich animal environments, enlightened animal care and caretaker devotion should all be made visible to the public within a setting which demonstrates the animals' innate competence. (p. 978)

The design movement to activate zoo animals may be compared to efforts to use architecture to combat obesity in people. "Active design" was developed to create spaces that encourage healthy lifestyles. Guidelines for active design were developed for the City of New York in 2006. Buildings that encourage activity often feature exposed stairways to encourage people to use them. Climbing stairs for just two minutes a day can prevent weight gain. Designers have developed prompts ("Burn calories, not electricity!") to encourage occupants of active office buildings. Offices are also being configured to relocate services a healthy walk away from the workplace. Convenience is giving way to movement. Shared office workspace achieves the same outcome. Natural light also encourages movement in office workers. Perhaps the most compelling active design feature is the provision of contiguous outdoor areas for walking and interacting with coworkers (Nasser 2012).

As Coe and Lee have concluded, the architecture creates opportunity and the enlightened, caring husbandry enables the use of space and the creative manipulanda within it (The manifesto that influenced the landscape immersion architects working today appears in Seidensticker and Doherty (1996) as "Jones' Principles".) The construct of welfare brings us full circle to the recognition that short-term intervention (behavioral management in its various forms) and long-term planning (exhibit design) is inextricably linked. No future master-plan should consist of only architecture as behavior must be a major factor in the design of future zoos and aquariums. Naturalistic habitats work even better when enrichment is added to the mix.

8.4 Rotating Animals and Habitat

The zoo design firm CLR significantly enhanced their reputation for innovation when they began to experiment with exhibit rotations (Fig. 8.3). They developed a "habitats-in-the-round" feature at the Louisville Zoo that revisited the experience of Apenheul in The Netherlands and the Columbus Zoo (Ohio), both of which benefited from the availability of a large population of gorillas. In a rotation concept, animals can visit multiple areas each day. The options provide different opportunities depending on the management preferences of keepers and curators. For example, a rotation facility could sequence predators and prey, or they could simply rotate groups of the same species and generate excitement through sensory enrichment or some other type of environmental intervention. We have experience in monitoring the pioneering rotation options at Zoo Atlanta, designed by CLR with the potential to rotate gorillas through four outdoor yards and contiguous indoor holding rooms. The idea was to introduce novelty and stimulus change, and to socially engineer a "perceived competition" between the many silverback males. It was not a great surprise that breeding commenced among the gorillas immediately upon their occupation of the new facilities. With pregnancies and later offspring to monitor, the staff never felt comfortable introducing the change represented by the rotation feature. It was just too successful too soon. However, some years later, Kristen Lukas evaluated rotation for her master's thesis at Georgia Tech. She discovered that activity levels and exploration increased when two troops were gradually transitioned to daily rotation in adjacent habitats (Lukas et al. 2003).

The exceptional rotation program at the Louisville Zoo with orangutans, siamangs, tapirs, babirusas, and tigers rotating among three outdoor displays and a large indoor dayroom, enabled keepers to randomize the order, sequence, duration, and timing of the animals' movement each day. It is a remarkable achievement. A formal study of this innovation, essentially a post-occupancy evaluation, confirmed the team's expectations:

> The results support the conclusion that exposure to varying exhibits produces variation in the behavior of the animals and elicits natural behaviors that would be unlikely to occur in a traditional single exhibit. (White et al. 2003)

Zoo innovators and welfare-oriented staff frequently encounter resistance to innovation just as landscape immersion was initially resisted in the 1970s. Visitors to the Woodland Park Zoo didn't immediately appreciate the fact that the landscape wasn't regularly mowed or trimmed; it was meant to be wild. Eventually, they warmed up to it, but change isn't immediately embraced. The next frontier may be the challenge of giving animals more control over the ambient environment. Today's technology could encourage animals to change internal lighting, activate moving air, deliver food or browse, or dial up their favorite auditory, visual, or sensory stimuli, for example televised images of wild chimpanzees, penguins skiing down a snow bank, or the delivery of favored culinary aromas. To render animals healthy and well, keepers and curators will still be accountable for activating the system, but the more that animals can control for themselves, the better. Although

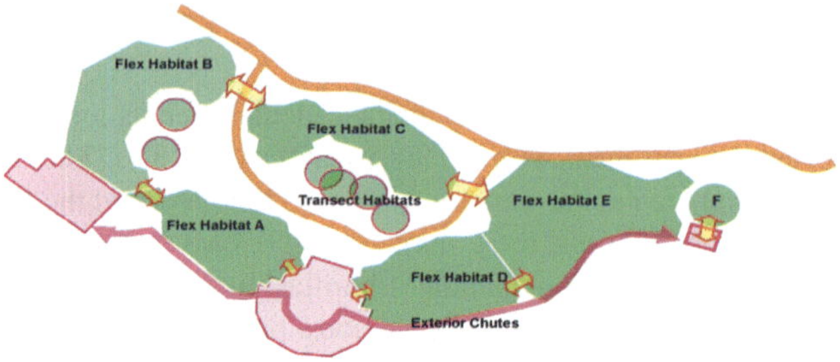

Fig. 8.3 Rotation concept represented by flexible habitats (G. Lee/CLR)

whatever the level of innovation, it is important to strive for objective evaluations. Architects have a name for the task; post-occupancy evaluation. Only a few zoos have engaged in a formal post-occupancy evaluation or POE (e.g. Maple and Finlay 1986; Finlay et al. 1988; Ogden et al. 1990, 1993; Chang et al. 1999; Maple 2005). Bronwyn and Ford (1991) issued a survey to 1,614 students in Australia who visited old and new gorilla exhibits at the Melbourne Zoo. The results strongly confirmed that the landscape immersion exhibit was superior to the barren enclosure in teaching about conservation. The investigators also reported that students developed more positive attitudes about the zoo after exposure to the new exhibit. A comprehensive POE for great ape facilities was recently published by Ross and Lukas (2006) based on a Lincoln Park Zoo renovation and expansion of its great ape facilities. The investigators discovered that zoo visitors spent 59 % more time within the naturalistic setting and moved more slowly through the exhibit compared to the old building. Visitors were also more considerate in the naturalistic facility as they exhibited significantly fewer instances of abusive glass-tapping. Because so many zoos open their new facilities with great fanfare and publicity, it takes awhile to really know what worked and didn't work. The best plans include resources to engage the evaluation prior to the animal moving into the new enclosure. It is important to obtain a behavioral baseline and if visitor behavior is a topic of concern, they should also be observed and/or interviewed prior to the change. POE's are essential if we are going to learn all we can about design that is intended to advance welfare. POE's have historically documented visitor attitudes about conservation. The next generation of research should provide an equal focus on attitudes about zoo animal welfare. POE's are also helpful when zoos are planning new exhibits and want to avoid overly stressful or violent social transitions. For managers of great apes, early studies of group formation at The Netherlands' Arnhem Zoo (van Hooff 1973) delivered useful information about chimpanzees adjusting to new naturalistic environments. An academic approach to zoo design requires an initial examination of what has been published prior to programming and ultimately design. The search for ideas, old and new, must be thorough to deliver true innovation. The empirical approach practiced at the Arnhem Zoo for so many years is a function of its close collaborative relationship to scholars and students at the nearby University of Utrecht. It surely helped that Anton van

Hooff, the older brother of Utrecht professor Jan van Hooff, was for many years the Director of Arnhem's Burgher Zoo.

8.5 Designing Big; Living Large

Zoos have been criticized for trying to do too much. We are still debating the optimal size and population for a zoo, but our critics have lobbied for zoos with a regional orientation specializing in the exhibition of animals found in the local ecosystems. David Hancocks strongly prefers this approach:

> I plead now for some immediate and major changes in zoos: first, a move away from exhibiting and interpreting only animals; second, better representation of small life forms, to give a more accurate view of the diversity and complexity of the planet's fauna; third, an attempt to show the functional roles that animals play in their ecosystems; fourth, closer attention to the total quality of life of all the animals in the collections; and, fifth, more regional specialization . . .

A good example of specialization is the Arizona-Sonora Desert Museum. This zoo has always provided a close look at small animals from the local biome, including 368 terrestrial invertebrates and 495 reptiles. Further, in the backdrop of a dense Sonoran forest of Saguaro cactus, *Carnegiea gigantea*, the desert ecosystem is easily viewed and interpreted in the only region in the world where this marvelous plant can be seen. A zoo's story is incomplete without botanical holdings and many zoos have rebranded themselves to function as a "zoo and botanical garden". The Desert Museum is zoo, botanical garden, and museum all rolled into one institution. Their mission statement is uniquely focused: ". . . to inspire people to live in harmony with the natural world by fostering love, appreciation, and understanding of the Sonoran desert." In botanically oriented zoos, endangered plants provide an opportunity to reach out to our local gardening communities, while the innate charisma of a Saguaro cactus cannot be disputed. Zoos and gardens with a healthy supply of living plants have even started to market odiferous species such as *Amorphophallus titanium*, also known as the "corpse flower," that blooms every 75 years, exuding the smell of rotting flesh. More than 10,000 visitors came to see the odd species (more than 6 ft. in height) during its exhibition at the Basel Botanical Garden in 2011. Another smelly plant from Indonesia, the durian (*Durio* spp.), produces a fruit that is a favored food of wild orangutans. The spiney durian makes an interesting educational artifact for zoo docents who must handle it with care. Many hotels and public places in Indonesia and Singapore have banned the plant from their premises due to its offensive odor. Wooden models can be obtained to show visitors the spiny external features of the plant. There is some inherent conflict in presenting animals among botanicals, as horticulture expert Don Jackson observed:

> . . . there is much disagreement among animal managers concerning philosophies of mutual coexistence between plants and animals within exhibits. Opinions range from the dedicated and determined plant 'protectionist' stance to the far opposite end of the spectrum. Frequently, funds designed for the protection of valuable mature trees within exhibits are

> cut from the construction budget. Exhibit construction itself results in considerable damage to trees . . . The design and construction of an efficient, cost-effective, and aesthetically pleasing structure that will, for example, protect a mature oak tree from the ravages of a full-grown male elephant is not a trivial task.

For major zoos and aquariums it is tempting to brag about the size of their collections (or populations as we prefer to call them). The Shedd Aquarium in Chicago, for example, exhibits 32,500 animals including whales, dolphins, lizards and turtles, and more than 19,000 fish. The Berlin Zoo boasts a total zoo population of 17,500 animals representing 1,500 distinct species. This is thought to be the most comprehensive population in any world zoo. The Georgia Aquarium, containing 8 million U.S. gallons (30,000,000 L) of marine and fresh water, is known as the world's largest aquarium, but is it good to be the biggest? Hoage (1996) commented on the history of this trend:

> The world's best zoo by definition had to exhibit the largest number of exotic species (especially the rarest or most recently discovered). Zoo animal collections at the time, it seems, were viewed as examples of "mother nature's" works of art. Obtaining a newly discovered Okapi was equivalent to the national art gallery's acquiring an original da Vinci. Collecting exotic animals became analogous to collecting rare or heretofore inaccessible art. (p. 136)

We have a catch-phrase that represents our idea of quality over quantity; "Fewer animals living large." In our view, the quality of life for exhibited animals is much more important than how many animals reside in the entire facility. "Living large" means living well. This is not an equivalent issue when you consider aquariums and zoos. Many aquatic species live in schools so presenting them in large numbers is normal and natural. Establishing the appropriate number of fish in enclosed space is both an art and a science, and a very delicate process. Even in a zoo, the right number varies with the species. To establish a breeding group of flamingos, a zoo needs at least twenty birds, as flamingos prefer living in close proximity in a large flock whether they are in the wild or in the zoo (Stevens and Pickett 1994). Future zoos will likely exhibit fewer species, but larger groups will be necessary, as we have seen in the trend to larger herds of elephants and larger groups of gorillas. A naturalistic simulation requires group size that is sufficient for natural behavior to unfold. The demand for sufficient space and larger groups led Hancocks to conclude that a few grand exhibits of elephants, placed strategically around the country, would enable zoo patrons to see elephants in a true naturalistic simulation, good for elephants and good for visitors, but not available in every zoo (Lemonick 2006).

Large breeding groups require much greater space and this has led to the development of open-range zoological parks. The first one was opened in 1931 in Bedfordshire, England by the Zoological Society of London. Known as Whipsnade Park, it covered 600 acres (2.4 km^2). The San Diego Zoo Safari Park has operated since 1970 on a 1,800 acre site (7 km^2). The Werribee Open Range Zoo in Melbourne, Australia occupies 500 acres (2.0 km^2). A major influence on the style if not the substance of these facilities was the early success of the "Lion Country Safari Parks" in the U.S.A. Only one of these wildlife theme parks remains, in West Palm Beach, Florida, but the opportunity to drive the family car through naturalistic displays of free-ranging African animals was popular for decades. However, with few exceptions safari parks were designed to attract and entertain visitors with an important but lesser focus on providing enhanced welfare. Of course, entertainment and welfare are not incompatible motives. Theoretically, even the most remote sanctuaries could be designed to encourage visitation.

An insidious tendency, after a natural habitat zoo is opened, is the conversion of green, open space to add more exhibits, a kind of urbanization process. It is the pressure to market new animals that leads to social/exhibit density problems. We believe that many zoos, and probably many aquariums, would benefit from a delicate "down-sizing" of the population to provide an enlarged living opportunity for the animals that remain. For this task, we like the business term "rightsizing". Today's competitive market forces challenge organizations to make decisions that are good for their customers and the animals. The market also dictates how we operate African safaris. People who pay large sums of money to visit Kenya must see the "big five" or they are disappointed. The safari operator often has to choose between quality and quantity in the limited time available to study and appreciate biodiversity. If we have to choose between exhibiting lions or leopards in a zoo, it is better to build a great lion exhibit than settle for mediocrity in both.

In designing for welfare, a generous amount of exhibit space encourages activity and exploration and ultimately leads to better health. For example, a new $12M elephant exhibit at the Honolulu Zoo, roughly ten times larger than their former quarters, resulted in significant weight loss in two female elephants after 5 months of living in the larger enclosure. Each animal lost about 300 pounds due to increased activity. To keep them moving and foraging, zoo keepers spread their food throughout the enclosure. Designing for welfare meant the installation of a massive scale so staff could weigh them daily and monitor their progress. In spite of the improved facilities and the successful activation of the elephants, an animal rights group, In Defense of Animals, was not impressed, rating the Honolulu Zoo habitat among the ten worst in the nation. Undeterred, Honolulu officials plan to study the elephant's use of space to determine its influence on their behavior while continuing their planned improvements.

A new Asian elephant exhibit planned for the Smithsonian's National Zoo reverses the zoo's original design which catered to the visitors rather than the elephants. The new exhibit, known as "Elephant Trails" features pools, sand pits, and a quarter-mile long forested walking trail where the herd can move along together. The principle design theme is the emphasis on a living herd of related individuals, a simulation of nature. The renovated indoor habitat is also built on a sand substrate and provides ample space for a social group. The indoor elephant house also deploys large pickle barrels hung from the ceiling to provide a source of vertical enrichment. This innovation stimulates natural exploration, play, and object manipulation. Although conservation is a major educational theme, the zoo has made a fundamental commitment to animal welfare with the focus of this innovative elephant facility.

One of the largest elephant facilities in North America is the emerging National Elephant Center (TNEC), taking shape on a 225-acre site in Central Florida, a project organized on behalf of the Association of Zoos and Aquariums, designed to serve the strategic interests of 73 collaborating zoos committed to breeding elephants. TNEC will have sufficient, flexible space to accommodate bull elephants, managed herds, and other social groupings. In another large project, the Birmingham (Alabama) Zoo is raising money for a 14 acre "Trails of Africa" elephant facility that will accommodate bull elephants, perhaps the most difficult design challenge for any zoo architect. However, many other North American zoos (Bronx, Chicago-Lincoln Park, Detroit, San Francisco, Santa Barbara, and Toronto) have decided to stop exhibiting elephants because the standards have changed so dramatically and the costs of compliance have risen so sharply. In 2006, Bronx Zoo officials announced that they planned to shut down the elephant exhibit after the death of one or more of the three remaining adult elephants at the zoo, and they will not replenish the herd. Other zoos worldwide have chosen to allocate the necessary resources to meet the new standards for exhibiting and managing elephants, but there is still debate and disagreement on whether we can ever achieve acceptable levels of welfare given the great size, complexity and psychological needs of elephants. Because so many zoos have elected to innovate with elephant exhibits, this collective effort can be regarded as a complex design experiment to determine how captivity can be shaped to influence welfare in elephants. In a few years, we should have some answers.

8.6 Encouraging Constructive Criticism

The value of focused workshops such as the Atlanta ethics workshop, the Detroit Welfare conference, and the Tufts elephant meeting is the opportunity to provide detailed reviews of key management, design, and exhibition issues. For example, participants in the Atlanta meeting agreed that AZA leadership needed to convene a meeting to unite zoo designers and animal welfare scientists to discuss the relationship between welfare and zoo design. In fact, meetings for this purpose resulted from independent action by leading zoos and universities. Reform often advances in an entrepreneurial direction. The Tufts publication *An Elephant in the Room* (Forthman et al. 2009) is an extremely important publication, the result of leading reformers pushing for change. Only a few AZA elephant-holding institutions were represented at this conference, demonstrating the inherent volatility of the elephant debate. At the time of the meeting, elephant holders and elephant critics were highly polarized and communication was strained. An example of the red-hot rhetoric is a statement from Jeff Williamson, former President of the Arizona Zoological Society, who wrote:

> In my view elephant exhibits in urban zoos – my own included – are failures . . . because they are woefully inadequate to the needs of the species. Elephants are intelligent, social and mobile creatures. We wrench them out of the wild in order to exploit them. We disrupt their herd structure and confine them in spaces that represent a tiny percentage of their daily range, then wonder why they develop chronic physical and behavioral problems: foot and leg trouble, arthritis, pacing, swaying and aggressiveness. (p. 226)

There is a long documented history of substandard elephant facilities throughout the world, so it is fitting to take a look at what experts believe are the absolute necessities for elephants, in essence the design and management details that will lead us to an outcome of optimal welfare for this species. To find evidence of excellence in newly constructed zoo exhibits it is helpful to first go online to ZooLex, a website devoted to the publication of exhibit ideas and technical details (including budgets) replete with photographs and diagrams. The brilliant concept of landscape architect Monica Fiby, ZooLex is registered in Vienna as a non-profit organization sponsored by the World Association of Zoos and Aquariums and other private donors. Its main objective is to improve living conditions for wild animals in captivity. With so many excellent historical examples of design excellence available for careful review, the ZooLex website is the best place for a design programmer to investigate design history in detail (Fiby and Petzold 2005).

Originally distributed as a white paper in 2005, revised in 2007, and subsequently published as an appendix to the Tufts publication ("Optimal Conditions for Captive Elephants") Kane et al. (2009) have written a detailed document that every zoo designer should read before they contemplate building an elephant facility. We can only review a few of the main principles. In their report, they observed that the ambient environment was an important variable that has been largely ignored by zoos. They recommended provisions to control heat loss in elephants and provide for thermoregulation when the temperature reaches dangerous limits. Elephants that cannot go out at night, for example, run the risk of overheating in a night house. Sensory variables are also important, so zoo designers need to figure out how to reduce noise in buildings made from concrete block, iron and steel. Crowd noises can also contribute to reduced welfare for elephants. Proper ventilation will buffer odors from toxic cleaning fluids or the build-up of urine and fecal material. Field workers agree with zoo experts that the zoo's feeding regime should resemble feeding in the wild, and elephants should live in complex social groups. Strong social bonds should not be broken if possible.

Kane and associates also identified occupational variables for elephants that included room to roam on a variety of substrates, variation in topography, and access to appropriate plant material or free provisions of cut fresh browse to encourage foraging. Access to dust, water and mud sufficient for submerged bathing and whole-body wallowing is also highly desirable in a zoo habitat devoted to keeping elephants occupied. Their recommendations can be summarized as follows:

1. Spacious quarters that permit foraging, exploration, and exercise; year round access to the outdoors, live vegetation, access to peers, and a reasonable degree of autonomy.
2. Freedom from overnight and other extended periods of chaining.
3. Lifelong protection of the natal bond between mothers and female calves, in the absence of extraordinary cause.
4. Freedom from dominance-based behavior management, including physical punishment, threat of punishment, isolation or deprivation.

Fig. 8.4 Massive indoor habitat for Asian elephants at the Cologne Zoo

Committed elephant exhibitors in North America, like the San Diego Zoo, have stepped up to the challenge of exhibiting larger herds. They understand that their facilities will be judged by whether they succeed or fail at bringing out the best in elephants. San Diego Zoo's 7.5 acre exhibit (Elephant Odyssey) features a massive 137,000 gallon pool for the zoo's seven Asian elephants. Five towering "utilitrees" were designed and built to provide an artificial source of shade and also deliver food operated by the animals. A similar structure has been built with 18 trellises (Supertrees) in a 250-acre Singapore water park development known as Gardens by the Bay (e.g. gardensbythebay.org.sg). An innovative "trunk wall" was also developed in San Diego to encourage keepers to reward elephants for demonstrating trunk dexterity. Large pools are enriching for elephants. The Basel Zoo in Switzerland promotes formal bathing events for their elephants and encourages their guests to observe them in the water. At Basel, the elephants are able to submerge into a naturalistic waterhole and they clearly enjoy the opportunity to play in water.

Zoo Atlanta opened a naturalistic elephant exhibit in 1989 when the three female African elephants were small. With advice from field biologists, the enclosure was equipped with a 40,000 gallon full-immersion pool. Although they could be induced to enter the pool and they clearly enjoyed it, the elephants preferred their designated mud wallow. Given the discovery that mud wallows were just as compelling as a deep water feature, the findings of Leighty et al. (2010) are puzzling. They documented territorial behavior in their pool where dominant elephants used the pool more often, but no such dominance was observed at the mud wallow. Immersive pools and mud both work for elephants, but effective design requires formal evaluation to discover how it works (Fig. 8.4).

The Association of Zoos and Aquariums continues to develop formal guidelines for their members who manage elephants. AZA recommends that institutions holding elephants keep herds of 6–12 animals. To exhibit larger herds, zoos will need to increase the size of their exhibits. In a recent study (Rees 2009a) of 495 Asian elephants and 336 African elephants living in 194 zoos in Europe and North America, the average group size was 4.28 animals. One fifth of the elephants lived alone (N-46) or with only one conspecific. The author recommended that zoos cooperate to build improved facilities and form larger herds. To this end, Oklahoma City Zoo recently opened the largest Asian elephant exhibit in the United States, nine and one-half acres in size. It includes three elephant yards, a demonstration pavilion, and a very large barn that some have called "a parking garage". The barn has eight stalls, a common area, and a sand substrate. Although the zoo currently houses only two female elephants, it has plenty of room for a herd. Outdoors the elephants can easily find shade under the towering wood pavilions where zoo keepers dangle food or objects to encourage natural foraging behavior. Old tree trunks are routinely provided and the animals have access to a massive pool. The staff developed the $13M exhibit with the goal of providing a "healthy, stimulating environment."

In a survey by Lewis et al. (2010), 89 % of the elephant-holding institutions in AZA committed to increase their holding capacity. In addition, 50 % of the institutions planned to reduce the amount of concrete in their enclosures while 41 % planned to increase the area of natural substrate in adjacent outdoor enclosures. According to this survey, 95 % of the facilities expected to have an elephant restraint device, and 80 % planned to increase the amount of exercise demanded from their herds.

8.7 Thinking Big About Rhinos and Hippos

Like the elephant, the black rhinoceros (*Diceros bicornis*) is understandably regarded as a "charismatic mega-vertebrate" with a lifestyle that is hard to duplicate in the zoo. Dr. Debra Forthman, who has contributed scores of superior ideas to the literature of elephant welfare, took a careful look at what she regarded as the primary issues when designing for rhinos and other ungulates. Forthman suggested spacious enclosures with long views to minimize startle responses when people suddenly appear. She also recommended that keepers retain dung piles during cleaning and novel scents be introduced to provide sensory enrichment. Like elephants, rhinos also need rubbing posts and both dry and wet wallows. Food and browse should be widely dispersed so the animal is prompted to move about the enclosure. Rhinos also benefit from opportunities to be solitary or social and from stratified enclosures with privacy areas hidden behind rockwork buffers. For designers, it pays to think like a rhino. Of course, since white rhinos (*Ceratotherium simum*) are grazers and black rhinos are browsers, these adaptations must be taken into account when designing their facilities. White rhinos are also more comfortable in groups. Asian rhinos are also specialized and present different design opportunities. Forthman's research applies to other large mammals that require thoughtful environmental design (Forthman et al. 1995; Forthman 1998).

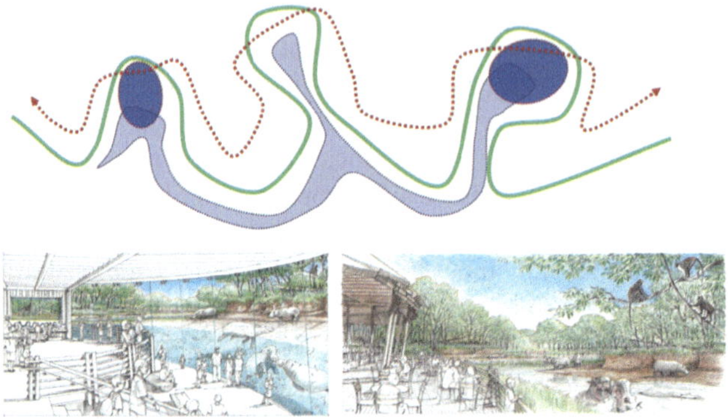

Fig. 8.5 A sequence of pools draining from higher to lower to activate hippos (Courtesy Gary Lee/ CLR)

Recently we began to study welfare standards for captive Nile hippos (*Hippopotamus amphibius*). The new standard for this species was pioneered at the Toledo Zoo with their Hippoquarium. With the acquisition of a high-tech filter, the water in hippo pools can be kept clean enough for underwater viewing of the animals; good for the hippos and a wonderful experience for zoo visitors. San Diego Zoo also built a riverine hippo facility with underwater viewing. However neither exhibit in itself activated hippos, and hippos confined at night are not visible unless the zoo is open. Even in the wild, hippos are active on the land at night, but not very active in the water during the day. One study that attempted to learn about hippo activity in a zoo was conducted at Disney's Animal Kingdom (Blowers et al. 2012). Nine female hippos were observed, the largest group of hippos in a North American zoo. The exhibit was riverine in shape, 121 m in length and 24 m in width. Hippos had access to water and land. The distribution of hippos at Disney's Animal Kingdom confirmed their preference to rest in shallow water. It is thought that hippos conserve energy by standing in shallower waters where they can easily raise their head to breathe, a pattern similar to that observed in manatees. The exhibit was large enough to provide both social opportunities and privacy throughout the day. In collaboration with zoo designer Gary Lee and University of California at Davis graduate student Kristen Denninger, we are trying to find new ways to activate hippos during the day. Lee's design, depicted in this chapter, provides for small pools in an elevated sequence with the water slowly draining from higher pools to the lower ones, until at the end of the day the hippos have moved from one spot to another in full view of the public (Fig. 8.5). Hippos benefit from this design by the necessity of moving about their enclosures to stay in contact with water. Of course, the ideal hippo exhibit would be expansive, permitting both day and night activation of the animals, and simulating their habits in the wild. Lee's design could also incorporate under water viewing, a feature more educational than welfare-based. Having experienced the joy and wonder of safari camps built around hippo pools in Africa, we recognize that a restaurant built around a hippo exhibit would provide an exciting zoo venue that would work for guests and hippos alike. Creative welfare-oriented and visitor-friendly exhibits are highly marketable. When the Walt Disney Company opened their innovative Disney's Animal Kingdom, with their impressive group of hippos, they missed a great opportunity to execute this idea as only Disney "Imagineers" can do. We hope they will eventually add this feature to the innovative terrain of the Disney Animal Kingdom in Orlando.

8.8 Simulations and Replications

After the success of Woodland Park's famed gorilla exhibit designed by Jones & Jones, a flurry of new ape exhibits were built throughout North America and beyond. Gorillas claimed the lion's share of the funding and the innovation, as zoos clamored to acquire the animals through the distribution process established in regions throughout the world. Although "welfare" was not an explicit design criterion, new gorilla exhibits demonstrated a growing awareness of the variables that really counted in exhibit design as enumerated in a growing scientific literature (Maple and Hoff 1982; Coe 1985). The most salient change was the effort to build gorilla exhibits large enough to contain family groups. Multiple gorilla groups were feasible at Bronx Zoo, Chicago's Lincoln Park Zoo, Cincinnati Zoo, Columbus Zoo, San Diego Zoo, and Zoo Atlanta, institutions that had more than one male and a sufficient number of females to form harems. When it opened in 1988, Zoo Atlanta exhibited three contiguous gorilla groups and one solitary silverback male. Seattle designers demonstrated in 1977 that groups of gorillas could be exhibited in grassy, planted habitats and this practice is now the prevailing naturalistic element in world zoos. A few older, indoor exhibits offer sufficient space but the occupants must recline on a hard, inflexible cement or gunite substrate that has to be softened with bales of wood-wool, shredded paper, or straw. If large exhibits are designed properly they also offer close-in views of the animals when they are utilizing shaded, cooled (or heated) feeding sites adjacent to large viewing windows. The animals make scheduled appearances to receive food rewards or just get comfortable. Sometimes the amiable gorillas, especially the younger ones, will interact across the glass with children. Exhibits planned for welfare make superior educational exhibits for people who appreciate the active, social, playful animals living an enriched life in a naturalistic habitat.

Some North American zoos that built indoor exhibits have later renovated them to give the animals' greater access to the outdoors. Gorillas in Atlanta are permitted access to the outdoors until the ambient temperature or wind chill index reaches 40 °F. In bad weather, when gorillas are confined indoors, very few people venture outdoors to visit the zoo, and the animals are protected from the weather in their heated night quarters. Cold-weather institutions must manage gorillas inside for many months and for this reason complex inner quarters need to be designed. As Jersey zoologist Jeremy Mallinson (personal communication) told us, "gorillas spend two-thirds of their life in their night house". It was his opinion that night houses needed to be upgraded to a much higher standard.

One of the best examples of the way that experience and science influence design is a paper by Jon Coe, Dwight Scott, and Kristen Lukas (2009) in which they outlined the variables that contributed to specialized facilities for bachelor groups of lowland gorillas. Their collaboration evolved from discussions at a meeting in the year 2000 on the management of bachelor gorillas. The special needs of all-male groups and their potential for dangerous aggression requires special attention by designers, as Coe and his associates demonstrated. The high productivity of scientists who monitored bachelor groups provided the evidence to drive programming and design (e.g. Stoinski et al. 2001; Stoinski et al. 2004; Kuhar et al. 2006).

Similarly, as a function of generous budgets, giant panda exhibits have been designed to provide naturalistic elements and encourage social behavior. An example is the David M. Rubenstein Family Giant Panda Habitat at the National Zoo. This exhibit is comprised of two adjacent yards with more than 12,000 sq. ft. of green space. It was designed to simulate the rocky, vegetated, and hilly terrain of central China. The naturalistic features of this functional exhibit include a grotto cooled by a system of pipes in the walls, low trees and shrubs that provide shade and cover, a fog grove creating a misty retreat from the heat, pools and streams for languishing in water, and rocks and fallen trees to stimulate climbing and exploration.

Panda exhibits are rare, only 16 zoos outside of China currently exhibit giant pandas worldwide, but all of the exhibits have attempted to generate innovations in animal welfare. In spite of this effort, only zoos in China have a sufficient population to approach optimal social opportunities. Even Mexico City's Chapultepec Zoo which enjoyed some early breeding success with giant pandas has resorted to artificial insemination in hopes of producing their first cub since 1990. Social enrichment and proper socialization for this species requires more than a pair of animals.

A recent article in the *New York Times* (Kaufman 2012) raised the issue of "mating choice" as a factor in breeding rare species. The article quoted Cheryl Asa who directs the Wildlife Contraception Center at the St. Louis Zoo. Dr. Asa wondered if captive breeding would be more successful if animals were given a choice of partners. When designers are pushed to develop facilities based on wellness or welfare, they immediately consider how a natural social system could be reconstructed in a zoo. Chinese biologists have succeeded beyond all expectations in their application of artificial insemination (AI) technology. After other world zoos could not breed pandas, they too began to utilize AI as a fail-safe method of producing offspring. Pandas have been produced through AI in Atlanta, the District of Columbia, and San Diego. However, China is still trying to use natural mating and recent successes have encouraged zoos outside China to breed first, and then implement AI. The brief window of opportunity for breeding this species makes the entire operation a delicate procedure. Redundancy makes it difficult to credit natural breeding with any confirmed success rate.

The next ambitious step in giant panda management is clearly the opportunity to operate overseas breeding/socialization centers with a small population of males and females in several institutions. Since panda breeding dyads have rarely succeeded without artificial insemination, we are not advancing the science of panda reproduction by avoiding social engineering. Indeed, for many endangered species, the next frontier in facilities design should be to design facilities with sufficient space and enough animals to enable socialization and mate selection. Currently only the breeding centers in Wolong and Chengdu have been able to engage in social manipulations of this kind. Breeding and socialization centers in other countries could also concentrate on resocializing and rehabilitating pandas that have not yet exhibited normal mating behavior. There appears to be a surplus of giant pandas in China and welfare is compromised for some of them. Therefore, overseas breeding centers with a behavioral mandate would improve the welfare of giant pandas in captivity. Behavioral management and supervised socialization has improved the behavior of chimpanzees and gorillas after a history of social deprivation. Rehabilitation is a numbers game and cannot be achieved in every zoo. Existing facilities for giant pandas could be renovated to hold additional animals, or new exhibitors can design their facility for this purpose in collaboration with Chinese institutions. The continuing success of panda breeding in China could result in over-crowding and even fewer animals that enjoy acceptable welfare conditions. The approach we are suggesting could become a high priority for the Chinese authorities, and it would change the current paradigm of emigrating only two giant pandas per institution.

A sufficient number of pandas will also lead to new discoveries. Chinese scientists have been innovators in reproductive technology but assisted reproduction is only one of many frontiers of discovery. Studies of giant panda behavior will lead to a better understanding of the panda's abilities and limitations including its sensory and cognitive skills. Graduate student research on giant pandas in Atlanta augmented by data gathered at the Chengdu Center for Giant Panda Breeding documented spatial memory and confirmed color vision in this species (Tarou et al. 2004; Kelling et al. 2006). Long-term studies of giant pandas will provide greater insight into individual differences and personality, variables that will likely influence their propensity to ultimately mate and parent successfully.

Problems with captive pandas mirror the historical challenge of breeding lowland gorillas. North American zoos confined gorillas in substandard facilities for decades, but constructive criticism from comparative psychologists, field biologists, architects, and reform minded zoo biologists provided the information that guided reform. The decision to form species-appropriate social groups, and innovations in exhibit technology reversed the trend and led to a population explosion. Managed populations of gorillas in North America and Europe are now thriving and meeting the expectations of the demographers who manage the data. Our reversal of fortune with lowland gorillas was nothing short of revolutionary and a model for managing other species including giant pandas (e.g. Maple et al. 2009). Chinese scientists and zoo officials are managing giant pandas to produce more animals that successfully mate. Once pandas are raising their own offspring, and socialization is normalized, AI may become an anachronism (Fig. 8.6).

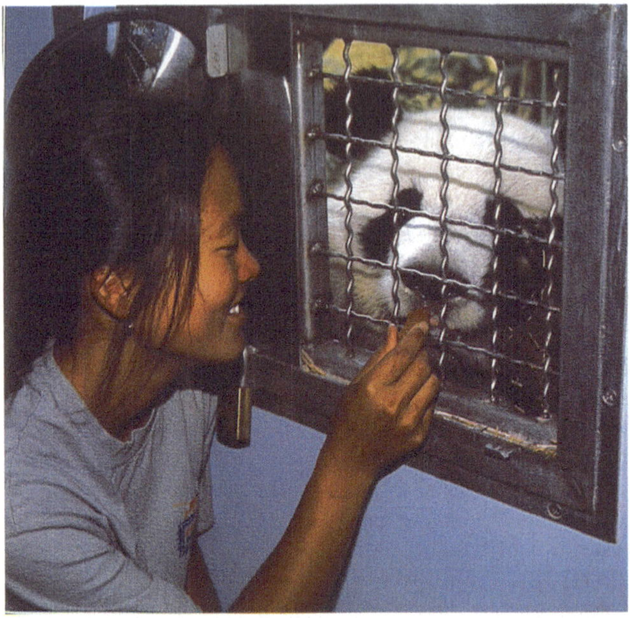

Fig. 8.6 Zoo Atlanta panda facilities were designed to enable research (A. Thompson)

8.9 Unleashing Natural Behavior

We experienced the dynamics of female choice when the senior author was Director of the Palm Beach Zoo. The zoo received two male Malayan tigers from the AZA Species Survival Plan (SSP) process in 2006 with the goal of eventually establishing a breeding pair. The two males were siblings born at the San Diego Zoo. Within a year after their arrival, an experienced female (Berapi) was obtained on breeding loan from the Bronx Zoo. Staff prepared her for socialization with both males but one of the animals was presumed to be the better candidate for breeding. Soon after her visual-olfactory introduction, safely carried out behind barriers, it was very clear that she had a different idea and expressed her strong preference for the other male. Of course, staff deferred and she happily bred. After an uneventful pregnancy, she delivered three healthy male offspring. Female choice does seem to matter to tigers, and this is one more dimension of welfare that should be considered as we design better zoos. We typically design facilities for a minimum number of animals, but this approach is beginning to change. Animals require sufficient social complexity to live a normal life, even if they are solitary or live in monogamous pairs. Therefore, facilities should provide additional holding space for other social partners or additional pairs. Such features, on exhibit and behind the scenes, will provide visitors a more authentic and compelling representation of the natural world, encourage rotation on the exhibit stage, and provide keepers the flexibility to manage social relationships. Good welfare requires sufficient numbers to operate

Fig. 8.7 Malayan tiger "Berapi" with one of three cubs born at the Palm Beach Zoo in Florida (K. Lovett)

a socially stimulating immersive environment (Fig. 8.7). As Asa et al. (2011) concluded:

> The major benefits from allowing choice are increased reproductive success, that is, higher birth and hatching rates and higher offspring survival, plus enhanced animal well-being. (p. 206).

8.10 When Experts Disagree

One of the most important books about zoos was written by David Hancocks (2001). Unfortunately, few zoo directors bothered to read the book, largely because Hancocks' strident criticism had alienated so many of them. His book should be required reading for all zoo directors and all students of zoo biology for its comprehensive compendium of history alone. Carefully researched, he introduces the reader to historic conversations between proponents and adversaries of the Hagenbeck school of open panoramas, reveals how imperious boards stifle the leadership of well-meaning zoo directors, and traces the zoo world's failure to fulfill its promise. In his book, Hancocks demonstrates the depth and breadth of his scholarship and his criticism. He is a formidable historian and passionate philosopher disappointed by the direction that zoos have taken. We've studied his work and regard him as an uncompromising idealist whose opinion must be taken into account in any discussion about the past, present, and future of zoos. So where is he right and where is Hancocks likely to be proved wrong? His objections filled a 200 page volume, but his early declarations are especially powerful:

> . . . when I lift images of zoos to mind, I find a jumble of unpleasant sights and sounds. Bored animals in small and sterile places . . . chain-link fences, trees made of epoxy resin . . . the echos of clanging steel doors as lions and tigers and bears are locked away for the night and the reverberating screams of chimpanzees ricocheting off bare walls . . . small birds in

> impoverished cages, snakes coiled on gravel, living in a green painted box . . . never able to stretch their body's length . . . steel feed dishes . . . tires hanging on chains, bounded with endless lumps of fake rock walls. (pp. xv-xvi)

His words paint a picture of hard architecture at the zoo, and Hancocks is absolutely right that hard zoos are no longer acceptable (Fig. 8.8). In fact, no zoo professionals that we know would endorse the kind of zoo that Hancock's deplores. They are anachronisms rapidly retreating into our memory of old and failed institutions. There are plenty of remnant menageries tolerated as "roadside attractions" and offering cheap thrills along the back roads of every nation. But Hancocks' isn't complaining about roadsides. He also believes that accredited zoos have failed to reach their full potential. He therefore wants to "un-invent" zoos as we know them and create a new kind of zoo, one that praises wild things, engenders respect for wildlife, and offers a holistic view of the natural world. Our colleague Jon Coe goes another step toward engaging the natural world, advocating a design concept he calls "un-zoos" that create opportunities to encounter and experience wild creatures at the edge of the wild. These are exciting possibilities but what do we do while we wait for the next revolution in zoo design? Furthermore, the "un-zoo" flies dangerously close to the flame of zoo abolitionists who recently solicited contributors to a conference under the banner of "transforming animal encounters in the twilight of the zoo." The call envisioned "novel modes of authentic encounter that might cultivate humanity's biophilic tendencies without degrading or abusing other animals." The convenors (Acampora 2007) dared to brand their effort "zootopian visions". If you are an idealist, it is frustrating to witness the slow pace of reform in the world's zoos and aquariums. But the pace of the industry inevitably quickens when leading zoos experience success in the marketplace. Welfare is a concept that may accelerate the pace of change as zoo- goers embrace the idea that zoo animals should live large and prosper. Meanwhile, as we await grand new exhibits designed with welfare in mind, our programs in behavioral management and environmental enrichment will provide a better life, and healthier, fitter population of zoo animals. With normal populations we have the opportunity to breed animals that will grow up normally in a social group. Managed breeding groups are not a trivial achievement as Colin Tudge (2001) asserted in his astute review of *A Different Nature*:

> Hancocks is wrong . . . to dismiss the role of captive breeding in modern conservation, and the contributions that zoos can make to it. Obviously, the ideal is to conserve wild animals in the wild. But this isn't always an option . . . Few zoos are ideal for captive breeding but . . . they must play their part . . . it is ridiculous to write off a strategy that is necessary for some because it cannot be applied to all. (p. 39)

In a germane opinion column in the *Sydney Morning Herald*, Hancocks (2007) made the case for the relevance of animal welfare to conservation. His belief in the superior morality of the welfare operating strategy leads him to the conclusion that welfare-oriented zoos would make better strategic decisions:

> . . . if zoos saw animal welfare as their central goal, they might become more effective conservation leaders. The exhibits, interpretation strategies, education programs, husbandry and collection would all be quite different in a zoo focused on welfare.

Fig. 8.8 Atlanta Zoo Kodiak bear exhibited on hot concrete with no shade, c. 1975 (T. Maple)

As Hancocks argues, a commitment to animal well-being ought to prevent zoos from exhibiting animals as complex as elephants. But is there no world zoo that could solve the problem of exhibiting, managing, and breeding elephants successfully? Do wild elephants live in such a perfect world that zoo professionals could not discover an effective zoo simulation and provide for optimal welfare? And what about Hancocks' suggestion that just a few "mega-exhibits" could be strategically placed in major cities, managed, and marketed as elephant destinations? Our most severe critics see the only ethical choice as the liberation of zoo elephants to live in sanctuaries, but there are many experts who are comfortable with the effective and comprehensive revitalization of elephant exhibits. California's Oakland Zoo has become a zoo animal welfare leader, in part because of their innovations with elephants. As they did with gorillas, zoo designers are working overtime to design and build optimal elephant facilities in zoos. One byproduct of the controversy is the elevation of elephants to the highest priority for facility upgrades. Dozens of zoos have already built new elephant exhibits, dramatically bigger and better, and many more are awaiting construction. We would not want to see zoo elephants living in anything less than facilities that achieve a new benchmark for welfare. With time and effort the exhibition of elephants is going to get a lot better.

We have faith in the brilliant designers who are planning future zoos and offering new solutions. Jon Coe's most recent efforts, in collaboration with his former partners at CLR, are working wonders at the Philadelphia Zoo (Matheson 2012). The architects have designed a structure that encourages monkeys, lemurs, and orangutans to use a set of enclosed trails to virtually leave the confines of their exhibit to explore the rest of the zoo. Zoo visitors see the animals in unexpected places throughout the zoo so it creates a sense of mystery and surprise. The zoo's chief operating officer, Andrew Baker, was ecstatic in praising the impact of this design innovation:

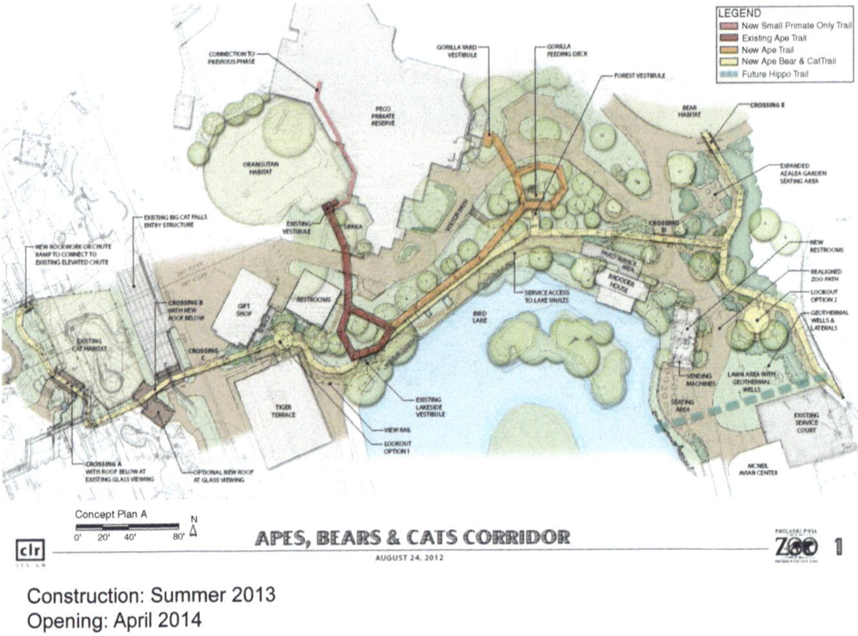

Fig. 8.9 Trail system enables movement outside enclosures at Philadelphia Zoo (Courtesy Andy Baker)

> We believe that the opportunity to travel, to explore, to choose to go toward things that are interesting, move away from things, {to} really control their own experience . . . is going to be incredibly enriching for the animals in our care.

The future of animal trails is even more promising as the zoo plans to add another 1,000 ft. to their "Treetop Trail" for small animals, and then use time-sharing principles to allow bears and great cats to alternate with orangutans on the Great Ape Trail. Eventually hippos, giraffe, and zebras will use a third trail currently in design. At the Tufts elephant symposium, we explored a similar idea for elephants with our introduction of the Buckhead Elephant Park concept, although this was an imaginary plan, so big (a greensward ten miles in length) that it would be nearly impossible to build (Maple et al. 2009). If elephants could literally leave the zoo at night and explore protected trails around the city, it would provide an entirely new definition for zoo animal welfare. Webcams would capture their adventures day and night and provide a new way to enjoy the zoo. Activity-based design and rotational systems, pioneered by Jon Coe and Gary Lee, have truly revolutionized zoo design based on the evolving principles of behavior and welfare and outside-the-box creativity. If Hancocks sees the modern zoo as half-empty, we see it as half-full (Fig. 8.9).

Curators and designers must also pay attention to small but frequently overlooked details. It is important to probe for hidden environmental variables that can compromise welfare; unobtrusive features such as potentially dangerous air pollutants, toxic

Fig. 8.10 Orangutan uses the trail system to emerge high above the exhibit (A. Baker)

cleaning solutions, loud noises, extreme temperatures, unnatural light, unusual fluctuations in weather or climate, or prolonged spatial intrusion by visitors. All of these frequently ignored factors can produce compromised welfare, especially in wildlife with sensory systems demonstrably different from our own. Some painful lessons have been learned in the construction of reptile facilities and aquariums when many delicate specimens were lost and facilities had to close to remove toxic materials. In 1966, the inadvertent use of formaldehyde in construction killed many animals in the new reptile house in Atlanta, resulting in a complete gutting of the facility and a one-year delay in opening the building (Desiderio 2000). Welfare begins with astute planning and design (Fig. 8.10).

Accredited zoos and aquariums have articulated a visionary agenda, but the process of change is inherently slow. Superior zoo animal welfare should be an easy sell for our visitors, our members, and our donors as they cannot enjoy zoos and aquariums that do not meet their expectations. They are likely to support any effort that clearly upgrades the quality of life for zoo animals. Great zoos are not built easily or quickly, but our intellectual capital is up to the task. At no time in the history of zoos have so many creative scientists, designers, and leaders been assembled to propel our institutions forward. Our next generation of zoos and aquariums will be unlike anything we have yet imagined.

Chapter 9
Launching Ethical Arks

If living creatures cannot be left in their original habitat, the least that can be done is to place them in natural and responsive surroundings—natural so that their character is not warped, and responsive so that their individuality and creativity are firmly respected.

Robert Sommer

Zoo and aquarium directors must be relentlessly vigilant to protect the core ethical values of the organization, but they must also stay focused on the task of keeping the ark afloat. For animal welfare to become our highest priority, a strong balance sheet will be necessary to support the investment. Make no mistake; welfare is an investment, and it will pay dividends if it "roots and shoots", as Jane Goodall might say. If there is continuing conflict between the ethical side of the house and the business side, there will be serious trouble down the line, so the business strategy has to become a welfare strategy. We believe that providing optimal welfare for zoo and aquarium animals adds value to the enterprise. Customers demand quality and they will not long patronize any organization that enables suffering or neglect. The priority of operating excellence starts with the health and welfare of every animal in the zoo. Active, fit, and healthy animals will also be a bigger draw and a source of satisfaction for the visiting public. Zoos and aquariums have sent strong messages about the primacy of wildlife conservation, and now we must elevate the priority of animal welfare by speaking clearly about our institutional commitment to superior animal care standards and practices.

A survey of zoo visitors in the United Kingdom by Reade and Waran (1996) revealed that 90 % of those surveyed expected animal welfare and environmental enrichment to be a zoo priority, and they expected to see evidence that these priorities were being implemented. The UK survey and decades of operating history support the position that zoo visitors, members, sponsors, donors, and staff actually want the animals to be the first priority of the zoo. If our customers are committed to the priority of wildlife, we should be eager to proclaim our commitment to zoo animal welfare. We believe that by advancing zoo animal welfare, we also strengthen our

T.L. Maple and B.M. Perdue, *Zoo Animal Welfare*, Animal Welfare,
DOI 10.1007/978-3-642-35955-2_9, © Springer-Verlag Berlin Heidelberg 2013

Fig. 9.1 Efforts to protect mountain gorillas may produce optimal welfare in the wild

commitment to wildlife conservation. By putting animals first, we acknowledge that the protection of wildlife is a continuum extending from the wild to the zoo (Fig. 9.1). Conservation and welfare are synergistic. Because so many zoogoer's report they visit zoos primarily for pleasure and family recreation (Fielder and Wheeler 1985), we must build our conservation and welfare platform in entertaining ways. An exciting raptor flight show demonstrates the close connection between the noble objectives of education, conservation, and welfare, and the sheer joy of witnessing free flight in hawks, owls, and eagles. The most jaded visitor will appreciate the opportunity to see active, healthy animals whenever they bring their family to the zoo.

9.1 Sustainable Science and Welfare

Research in ecological economics demonstrates that certain segments of the general population are willing to give up other goods and services to improve the well-being of another species (Naald and Cameron 2011). This finding is based on a study that examined the willingness of consumers to pay more for eggs provided by free-ranging rather than cage-bound chickens. For most of us, this is a moral imperative but it can be seen from the standpoint of cost-benefit analysis and defended on that basis. This is an important concept for zoo directors who must calculate their institution's capacity to raise money on an animal welfare platform. In our opinion, fund-raising for conservation is also predicated on the welfare of wild animals. They suffer from habitat encroachment, poaching, disease, and natural and man-made disasters, even civil warfare that compromise the living standards of wild animals. As we make zoos better for the wildlife we exhibit, the price of the experience may go up. We believe that zoo visitors will be willing to pay a higher price if it is unequivocally linked to tangible improvements in the living standards of the animals they want to see. Many zoos have also demonstrated that visitors willingly make spontaneous donations to conservation projects when prompted in the right way.

The World Association of Zoos and Aquariums confirmed its organizational commitment to animal welfare in the first "World Conservation Strategy" published in 1993. These superior practices and standards are based on an emerging science of zoo animal welfare, and disseminated through the financial support of institutes and foundations dedicated to the improvement of conditions in captivity. The world's best zoos rely on evidence to make ethical decisions benefitting the animals in their care. Although Hediger suggested long ago that "science was always last in the zoological garden," the modern zoo depends on scientific data to make the case and advance the cause of animal welfare. We therefore conclude that science cannot possibly be considered the last priority of any modern zoo or aquarium. Even smaller institutions recognize the importance of scientific partnerships to elevate their participation in the science of zoo biology and animal welfare. There is no zoo so isolated that it cannot find an effective and willing scientific partner. Although we have made significant progress in recruiting scientific zoo biologists and we have strengthened the role of science in zoos and aquariums, and despite the existence of several highly regarded scientific zoos, as a profession we are still relative underachievers in the applications of science and technology. Our business plans should reflect a greater investment in science by each and every zoo, and zoo executives need to make an effort to educate themselves and their boards on the value and importance of smart, evidence-based decision making, and the consideration of new ideas. To make a difference in species and ecosystem survival while simultaneously advancing health and wellness, astute zoo directors recognize the need to allocate sufficient resources to both field conservation and zoo animal welfare. As Anderson et al. (2010) clearly demonstrated, scientific zoo biology cannot succeed without the strong support of the Chief Executive and the institution's Board of Directors.

The philosopher Dale Jamieson (1985) asserted that zoos and aquariums can only be justified by their contributions to public education. We are therefore ethically obligated to provide the highest standards of living for the animals that provide this noble service to humankind. Ethical arks are evolving and exerting their dynamic leadership throughout the world, but how can we expand the reach of this trend? One way is to keep animal welfare visible and vital at workshops, national and international conferences, and in collaborative local programs. Major symposia have been published in peer-reviewed journals, and centers and institutes dedicated to welfare are now located in major institutions in Australia, Europe and North America. These centers have significantly raised the public profile of animal welfare. Leading universities throughout the world have also hosted benchmark lectures and symposia where the science of animal welfare has been dissected and debated by academics. Important public events, conferences, and workshops have been supported with funding from diverse sources such as the Institute for Museum and Library Services, the National Science Foundation, Morris Animal Foundation, Universities Federation of Animal Welfare, the Association of Zoos and Aquariums, and many other private and governmental sources. In Europe, the Paignton Zoo in Devon (UK) has hosted a sequence of design conferences showcasing the world's leading experts in zoo architecture and zoo biology to identify and discuss new trends and new ideas that benefit animals (e.g. Plowman and Tonge 2005).

For animal welfare science to become animal welfare practice, zoos and aquariums must step up to the challenge of growing our intellectual capital. Outside resources to support the institutional initiatives of curators, keepers, and veterinarians are critical as there aren't enough scientists to advance zoo animal welfare by their efforts alone. In marketing the cause of animal welfare, we should focus on the dedicated zoo keepers and other personnel who work so closely with the animals they love. Our critics have no idea how much zoo professionals care about the animals they serve on a daily basis, and these personal stories must be shared with the public. This was the sentiment of Rabb and Saunders (2005) when they wrote:

> . . . to succeed as conservation centers, zoos and aquariums need a much better understanding of people's fundamental psychological relations to animals and nature . . . we believe that it is essential for us to foster caring concerns and caring behavior for animals and nature if we are to stay in business, and if we are to carry out the world conservation strategy of *Caring for the Earth.* (p. 2)

9.2 Leaders in the Midst

The philanthropist and media entrepreneur Ted Turner is fond of the expression: "Lead, follow, or get out of the way!" Now is the time for strong leadership and the right moment to share what we know. Increasingly, leaders from highly successful zoos and aquariums are exporting their ideas and techniques to developing institutions in Africa, Asia, South America, and throughout the world. Like conservation, the welfare cause must become a global priority. Technical help is needed as zoos in Thailand, Indonesia, Malaysia, and Singapore, for example, have all been criticized for animal welfare and illegal trade violations. One suggestion from regional experts calls for wealthy zoos to adopt poor institutions in Southeast Asia to alleviate animal suffering, upgrade operating standards, and promote conservation (Agoramoorthy 2004). Regional zoo associations such as the Southeast Asian Zoo Association (SEAZA) appear eager to strengthen their standards and practices. SEAZA's new Ethics and Welfare Committee is now engaged in evaluations of member zoos to identify animals in distress and to identify unsatisfactory living conditions. This is a positive step but the challenge should not be underestimated. Reform will require diplomacy and patience, while exporting significant change will be costly. We must acknowledge that there are a number of superior zoos in Asia and in Australia that can also provide leadership to smaller institutions in need of financial support and mentoring. Formidable travel distances will also limit the extent of the change. All responsible zoos and aquariums, throughout the world, should work together to provide assistance to any institution where animals are suffering. The challenge extends to literally hundreds of marginal institutions (roadside attractions) in North America that somehow continue to operate despite the scrutiny of governments and humane organizations.

By partnering with universities and with humane societies and other animal welfare organizations, we can collectively advance animal welfare on many fronts. For years, we have trained young professionals and transferred conservation technology to developing countries, largely because these sites are where conservation must make a stand. We should also recruit young professionals, students, and government bureaucrats from distant nations to study animal welfare science and practices at the best world universities and affiliated centers and institutes. Funding from the west should be deployed to mentor partners in both conservation and animal welfare and to bridge gaps between them. Indeed, we can launch an armada of ethical arks if we activate all of our opportunities for collaboration with friendly animal welfare organizations. Reaching out to help local humane societies to offer adoptions of domestic dogs and cats is an easily arranged event for most of us. Detroit Zoo has provided their popular and accessible venue for this purpose for many years. Imagine the good will generated if an event such as this was national or international in scope. Humane societies, zoos, and aquariums have good reason to collaborate. Kreger and Hutchins (2010) envisioned other opportunities to exert leadership in animal welfare:

> . . . zoos could provide emergency services to non-zoo animals. Animal care staff can be promoted as animal welfare experts. Many zoos dispatch staff to help rehabilitate wild animals affected by oil spills. Aquariums rescue stranded marine mammals . . . They can provide advice for care of pets or care and rehabilitation of local wildlife. If they cannot temporarily maintain injured local wildlife or unwanted exotic pets, they can provide contact information for those who need it. Zoos can also partner with wildlife sanctuaries and rehabilitation centers to provide technical assistance or adopt non-releasable animals . . . (p. 8)

Indeed, Sea World parks have demonstrated just how effective our institutions can be in the rescue of injured wildlife. Spanning four decades, Sea World specialists have rescued more than 18,000 animals in their collaborations with local, state, and federal wildlife agencies. The organization aims to rehabilitate all rescued animals for a successful return to the wild. When this cannot be accomplished the animals are provided lifelong care in Sea World parks. Sea World veterinarians have been particularly effective in rescuing Florida manatees injured by recreational boaters or suffering from exposure to unusually cold waters. Other Florida institutions including Disney's Living Seas Aquarium, The Florida Aquarium, and Tampa's Lowry Park Zoo have also participated in labor-intensive, expensive manatee rescue operations.

Modern zoos are largely self-sustaining. We have worked hard to establish breeding populations that are carefully managed to be sustainable in the long-term. Occasionally, animals are rescued to live in zoos; orphaned grizzly bears, rogue polar bears, injured manatees and bald eagles, and confiscated chimpanzees are examples. When this happens, zoos occupy the moral high ground, and it usually occurs in partnership with governments or with humane organizational partners. In the case of the North

American condor breeding program, the plan to capture all remaining California condors to facilitate a zoo-based breeding program created controversy. In the end, all parties agreed that captive breeding was the only way to prevent extinction of this species. By all accounts, the breeding program has been very successful, and zoo-bred condors are now flying free again. As this book is being written, another controversy has generated debate in the *New York Times* and other major media. A group of North American aquariums has applied to the U.S. National Marine Fisheries Service for a permit to import 18 wild-caught belugas (*Delphinapterus leucas*) from Russia. Because this action would be the first time in two decades that the United States government has allowed importations of marine mammals purposefully captured for display, critics have argued that it dramatically expands the market for cetaceans. According to officials of the Humane Society of the United States, an average of 21 belugas have been captured in the Sea of Okhotsk every year since the year 2000, totaling more than 250 individuals (HSUS 2012a). Other countries, including China, have created this market, and the sudden entry of American institutions as recipients of wild cetaceans is troubling. The aquarium consortium argues that the captive population of belugas is not self-sustaining and new genetic material is needed to strengthen the captive population. As the AZA is supporting this application, zoos and aquariums collectively are once again on a collision course with the HSUS. From a welfare perspective, capture in the open seas is stressful and dangerously disruptive to the well-being of these advanced creatures. It will not be easy to justify this action on the basis of conservation or welfare, although the educational value and emotional appeal of belugas is unquestioned. Unlike previous acquisitions of belugas, this one is not a rescue, although the animals may be living in a distressed condition as they await their eventual translocation. In our opinion, it would be a better practice to thoroughly debate a departure from the norm before it becomes a media controversy. The acquisition of these whales is not illegal, but is it ethical?

To reach its full potential, the ethical ark will require strong hands at the helm. Zoo directors will need more than business acumen to steer ethical arks through a sea of criticism from media hysterics and animal rights extremists. To stay the course in reforming zoos and aquariums, zoo directors will have to demonstrate their commitment and exercise their leadership each and every day. In the end, if visitors feel sorry for animals in the zoo, we have failed to meet the challenge of humane exhibition. The zoo must become an ethical, caring, uplifting oasis where wildlife can fulfill their destiny as self-sustaining ambassadors for the natural world. Dedicated, effective, and worldly leaders will be required to achieve this noble outcome.

9.3 Strange Bedfellows

Partnerships with humane organizations can take many forms. Often the relationships are transitory but their impact can be significant. One of Zoo Atlanta's first experiences in working with a humane organization concerned the rescue of the gorilla Ivan, an animal that had lived in solitary confinement in a 14 × 14 concrete

Fig. 9.2 The rescued lowland gorilla Ivan in his prime at Zoo Atlanta (J. Sebo)

and steel cell for 27 years in a department store in Tacoma, Washington. Although he was loved by all who knew him, his cage was uncomfortable, barren, and demeaning. He had been isolated from his own kind since his capture in Africa, and he was humanized to an extreme. When the Progressive Animal Welfare Society (PAWS) began to lobby for his release to a zoo, Zoo Atlanta was asked if it could re-socialize him in its new state-of-the-art Ford African Rainforest facilities. PAWS then entered into a lengthy litigation to liberate him from the restricted department store environment. The negotiations were difficult but Ivan was eventually moved to Atlanta and introduced to two female gorillas. For nearly two decades he lived in a naturalistic immersion habitat, with access to the outdoors, other gorilla social partners, and friendly keepers. The vibrant and complex naturalism in the Atlanta exhibit served without question as an upgrade in welfare. He copulated with a gorilla only once indicating that, unlike Willie B., he didn't make a full and complete recovery. To the experienced observer, Ivan looked completely normal. Although Ivan was famous in Tacoma, Zoo Atlanta didn't acquire him for his box office appeal; it rescued him. He died on August 19, 2012, just short of his 50th birthday, a long life for a lowland gorilla. As roadside attractions continue to fail due to lack of visitor interest or legal entanglements, more rescues of exotic animals by accredited zoos may be necessary (Fig. 9.2).

Soon after Ivan had been safely relocated in the Ford African Rainforest exhibit in Atlanta, the world became aware of the plight of six polar bears marooned in a travelling Mexican circus. One of the polar bears was alleged to be a Zoo Atlanta bear, the offspring of a pairing that produced two baby bears in the early 80's, prior to the senior author's tenure as Chief Executive Officer which began in 1984. One of the bears, a female, was sold to a German zoo and died there. The other, a male, was sent to the San Francisco Zoo by the senior author in 1984, a decision predicated on the substandard facilities in Atlanta and approved by the Mayor. "Andy" bear was named after Mayor Andrew Young. A representative from People for the Ethical Treatment of Animals (PETA) contacted us in 2001 with information that the circus claimed it had legally obtained the bear from the Atlanta Zoo. An examination of blood from the parents, the known offspring in San Francisco, and a comparison to blood samples taken from the bear in the circus, clearly demonstrated that the circus bear was not a match to the DNA of Andy. Although PETA wanted to confiscate the suspect bear on grounds of cruelty, we argued that the circus should be cited for fraudulent identity theft. In the dead of night, federal officers of the U.S. Fish and Wildlife Service, in cooperation with AZA bear experts, removed the animal from its night quarters, crated it, and flew it out of Puerto Rico to Baltimore, MD. This unusual bear repossession required cooperation from partners with an adversarial history, but led to PETA's sheepish acknowledgment that this bear and the others, confiscated later, were better off in the AZA zoos that agreed to receive them. The U.S. Fish & Wildlife Service honored Federal Express for its important role in crating and flying the animals to safety. Other vivid and dynamic rescues like these will occur more frequently if we are serious about exotic animal welfare and recognize the need to extricate bears, cats, snakes, monkeys and apes living in deplorable conditions. A very similar situation currently prevails in India where the federal government has decreed that lions, tigers, leopards and primates can no longer perform in circuses and must be retrieved and rehabilitated by Indian zoos administered by the Central Zoo Authority in New Delhi (Gupta and Chakraborty 2005). If this trend continues, ethical arks will have to consider an expanded role in the rescue of exotics confiscated by governments and advocated by humane organizations. We are the only credible organizations with the facilities and expertise to help. This problem will be particularly acute in regions where private ownership of exotics is out of control. While there are only 5,000 tigers remaining in the wild, it is estimated that there are more than 15,000 tigers in backyard cages and dilapidated roadside attractions.

9.4 "Be the Change!"—Gandhi

Business ethics is enjoying a renaissance in schools of business in response to an epidemic of well-publicized corporate transgressions. Non-profit business practices have also generated media scrutiny in recent years. As zoo and aquarium boards examine their paperwork for conflicts of interest and other ethical concerns, they should also take the time to strengthen institutional support for animal welfare.

Certainly zoo animal welfare should be a top priority in every strategic plan and a highlight of every annual report. Management should actively monitor the welfare of every individual and every species and regularly report its findings and concerns to the board of directors. If we elevate its visibility and its priority, the welfare of zoo animals is bound to improve. The institutional mission statement must articulate and affirm core values that support animal care and welfare. The commitment to animal welfare should not be hidden or oblique. Just as conservation has become a salient feature of the modern zoo, so must animal welfare.

In one generation there has been substantial, even revolutionary change in zoos. In today's leading zoological parks, open, barrier-free exhibitory is the norm, not the exception. Our institutions prefer to configure species-appropriate, complex social groups, and often achieve mixed species exhibits within larger and extremely accurate landscape simulations of the natural world. Animals are expected to raise their own offspring so future generations will also engage in normal, natural parenting. We recognize the distinct individuality of animals and understand that they need social as well as intellectual stimulation to be healthy and well. The modern zoo and aquarium requires thoughtful intervention by caring, intelligent keepers, curators, and veterinarians. Only the highest standards of medicine are practiced in the world's accredited zoos. The end-product of our creative exhibits, intensive husbandry, enrichment and training programs, will be animals that are active and fit, socially competent, and interesting to observe. The ethical ark is both educational and inspirational, and highly marketable. By introducing our guests and members to the issue of zoo animal welfare, we can extend their concern to the health and welfare of all living things.

In this book we have consistently advocated collaboration with academics, including philosophers and environmental ethicists such as Bryan Norton, the lead editor of *Ethics on the Ark*. To advance the standards, practices, and operating principles of ethical zoos and aquariums, we need to give consideration to every good idea. Some of these ideas will come from intellectuals outside the zoo. In a new book, *Ethics and Animals*, by Lori Gruen, the concept of "wild dignity" was addressed. As Gruen argued: ". . . when we prevent them from controlling their own lives, we deny them their wild dignity. In contrast, we dignify the wildness of other animals when we respect their behaviors as meaningful to them and recognize that their lives are theirs to live."(p. 155) Most of our work to date honors this idea. We want zoo animals to be autonomous and in control of events in their life. We want them to select their mates, choose from a variety of food items, live in a complex environment, and—for many of them—spend the night outdoors. The trend is clear; the ethical ark is in favor of as much wild dignity as we can safely provide. We need to continue to challenge our assumptions about zoo animal welfare, and extend its boundaries to encompass greater freedom and dignity. We can accomplish more on behalf of the animals if we are receptive to new ideas and innovations regardless of the source.

Zoos and aquariums are in danger of becoming mere interpreters of science rather than organizations that generate new information. It is essential that we develop a sophisticated understanding of the biology and the behavior of the animals we manage. Our access to these unique life forms requires a commitment to learn from them. If we cannot demonstrate our expertise in the realm of science, how can we be trusted to make good decisions on their behalf? There is so much soft science being disseminated in the public media and through our own social media that our credibility is threatened. The authority to acquire, manage, and study wild animals is a function of our expertise and our reputation. For this reason alone zoos and aquariums need to recruit highly educated and well-trained scientists with the capability to conduct original, competitively funded research, provide accurate information through public education and institutional messaging systems, and help guide responsible decisions that engender public trust. Lacking such expertise, as dedicated professional staff or in strategic research partnerships, we are vulnerable to adversarial criticism and more likely to make serious mistakes in judgment. To be successful centers of public information and to speak out on behalf of wildlife, zoo and aquarium professionals must be well prepared and they must be entirely believable.

9.5 Conservation with Utmost Care

A new development in animal welfare is known as "compassionate conservation". In a column published in the British journal *New Scientist*, Mark Bekoff (2010) observed that in the practice of conservation we can unintentionally cause animals to suffer. One disturbing example is the translocation of wolves to restore the Yellowstone National Park ecosystem. In one area of Yellowstone more than 90 % of resident coyotes have been killed since the reintroduction of wolves. If individual lives matter then this reintroduction can be debated, even if the intent of wolf translocation is to restore a historical balance of predators and prey. Clearly, in the absence of wolves, the robust coyote population expanded to fill the void. Similarly, while compassion would dictate that all birds damaged by oil should be cared for, not just the endangered species, it also supports the proposition that introduced aliens such as Burmese pythons (*Python molurus bivattatus*) deserve a measure of humane treatment as they are systematically removed from the Everglades. This will not be a popular idea in Florida, as there may be literally thousands of mature pythons propagating in the ecosystem and they will surely degrade the welfare and threaten the survival of indigenous Florida wildlife. Compassionate conservation is not yet regarded as conventional wisdom, but Bekoff articulates the guiding principles of the movement as follows; do no intentional harm; respect all life; treat all individuals with respect and dignity; and tread lightly when stepping into the lives of animals. This is a fitting credo for ethical arks committed to wildlife and ecosystem conservation and animal welfare.

Zoos and aquariums have been traditionally organized to provide education, conservation, recreation, and research services, but these provisions are interrelated and multi-dimensional. The value of zoo medical experience in the field is well documented, but field experience can feed back to the zoo as well. A recent study by Dierenfeld (2006) and her colleagues described details about the consumption of browse in rare Sumatran rhinos. As delicate as this species has proved to be in captivity, their nutritional research in the range country should be very useful to zoo biologists who are trying to establish a captive population. One could easily argue that the optimal welfare of zoo rhinos, and other creatures, depends on an active agenda of conservation and research in the field.

It has taken decades to firmly establish conservation as the primary purpose of a zoo, and yet our critics dare to question the strength of our commitment. With the exception of the Bronx Zoo and a few others, we don't spend enough money nor do we have the impact of mission-focused field conservation organizations like Conservation International, the Nature Conservancy, or the World Wildlife Fund. However, the collective commitment to conservation by organized zoos and aquariums is impressive and growing (estimated $130 M for 1,970 conservation initiatives; AZA 2010 data), and our global partnerships can make a difference for many vulnerable species and ecosystems. However, only a few zoos spend 5–10 % annually on conservation. By contrast, in every zoo, the annual budget for health and welfare (essentially medicine and animal care) can be more than 50 % of the entire operating budget, and by virtue of allocated dollars, clearly confirms the priority of zoo animal welfare. Even if we agree that zoo animal welfare and wildlife conservation are co-equal priorities or that welfare is a close second to conservation, our daily commitment and focus on animal care dwarfs all other responsibilities in time and money. Because so many zoos fail to acknowledge that animal care is actually a subset of animal welfare, we don't receive credit where credit is due. It is the thesis of this book that by our history and by our intent, animal welfare is the primary enabler of our bond with the animals entrusted to our care. By honing our skills at managing animals in the zoo we are preparing for utilizing these same skills as managers of the wild. Conway (2011) made this point clear:

> Virtually all wild animal populations restricted to reserves will eventually need some level of curatorial care, such as population management and scientific research as well as greatly heightened veterinary support and traditional protections. Their ultimate preservation may ultimately entail such zoo tools as accreditation, International Species Information System, and even Species Survival Plans, and zoos and parks should both be employing such tools as risk and population viability assessment.

Our conservation partnerships are increasingly welfare based, and our institutional skills can be taught and replicated just as we transfer technology to other nations. Zoos must become comfortable with an organization that can do justice to both conservation and animal welfare, carefully coordinated themes that require investments to meet the challenge of saving animals from suffering and from extinction. An example is the Australian Koala (Fig. 9.3) which suffers from diseases and the danger of wildfires. Medical work in zoos is contributing to their survival in the wild. Conway also champions specialization, describing the great success of specialized avian programs such as the International Crane Foundation, Peregrine Fund, and the Wildfowl & Wetlands Trust. Unfortunately, traditional zoo bird collections have proved incompatible with sustainable propagation or superior care. Conway and Hancocks share the view that specialization is the better approach to planning a zoo. Specialization serves the demands of conservation and welfare.

The zoo that specializes in rare fauna, invests for successful propagation and health management, forms partnerships in range countries to support parklands and

Fig. 9.3 Zoos are contributing knowledge to achieve optimal health and welfare in koalas, a delicate species in captivity

managed ecosystems, and recruits public support to help sustain these partnerships makes an ethical commitment to wildlife. Ultimately welfare and conservation are compatible objectives that work well together. Our visitors and our donors expect nothing less than our best efforts to keep zoo animals healthy and well while contributing to a growing population of genetically diverse priority species. Institutions that achieve superior standards and practices in animal welfare do not detract from our shared conservation mission (Fig. 9.4). It is essential that we admit to our allegiance to both causes. Like wildlife conservation, animal welfare must be measured by deeds not words.

At the St. Louis meeting of WAZA in 2009, Russ Mittermeier of Conservation International advised the assembled zoo directors to embrace the role of promoting global conservation by exhibiting and promoting keystone endangered species; Malayan tigers, Sumatran orangutans, California condors, giant pandas, Queensland koalas and the like.

Conservation education is our specialty, and field conservation organizations understood that this was the most important contribution zoos could make to the conservation movement. To continue doing this well, we must ensure that the charismatic mega-fauna who represent their wild kin are presented in a realistic and wholly inspirational context. Animal fitness and health are essential features of exhibits that teach us about wildlife conservation and they must contribute to the sustainability of wild populations. Only active animals, behaving naturally, can achieve the results that our conservation allies expect. Clearly, animal welfare and conservation education are linked in the zoo.

Zoo biologists at the Lester E. Fisher Center for the Study and Conservation of Apes have taken seriously the challenge to generate support for conservation. In an article in the journal *Science*, Ross et al. (2008) called attention to the negative effects of television, print advertisements, and movies that portrayed chimpanzees as "frivolous sub-humans," rather than highly intelligent and endangered primates in need of protection. The Lincoln Park scientists suggested that changes in the image of

Fig. 9.4 Dr. Michele Miller and field veterinarians in Africa whose work demonstrates the close kinship between welfare and conservation medicine. (Courtesy M. Miller)

chimpanzees will likely generate positive public interest, respect, and commitments to conservation that will benefit all great apes. Of course, zoos work hard to present positive images of all animals as their welfare depends on the loyal support of the public. Allies in the entertainment industry must be converted to share this concern. Lincoln Park's innovative education program "Project ChimpCare" as directed by Dr. Ross aims to develop a strategic vision for the captive chimpanzee population in the United States so that sustainable housing and appropriate care can be provided for every chimpanzee.

While our critics are demanding better living standards for zoo animals, they also want us to lead the charge to protect animals in other settings; the horses that pull buggies in New York City, sharks awaiting slaughter as an ingredient in soup, a harassed and harpooned nation of whales. How much leadership in animal welfare can we accept without running the risk of losing our focus and our direction? As we concentrate on our primary responsibility to the animals we care for at the zoo and in the natural world, our ability to serve as humane and creative stewards of exotic wildlife will surely inspire reform in other venues. Speak out we must, but selectively and carefully. We must wisely choose our partners as we venture into an uncertain media environment. Our financial and human resources will always constrain our consciences, and we'll be hard pressed to accomplish more than our core supporters empower or expect. By continuing our conversation with the responsible animal welfare community we can play our most productive role. Our empirical approach to welfare is portable, and we should continue to practice evidence-based animal welfare in every situation where our services are rendered, and demand the same objectivity from others.

While accredited zoos and aquariums hone their welfare agenda, species by species, exhibit by exhibit, zoo and aquarium associations must also advance their credentials and their commitment to zoo animal welfare. The good reputation of an association should be deployed as a platform to lead rather than follow the lead of others. Consistent, coordinated messaging will prevent unproductive debate and provide for a common strategic approach to welfare. So far, a critical mass of national and regional associations in Australia and South East Asia seem to be moving in the same direction. It is essential that all zoos be held to account for the welfare and well-being of the animals they responsibly manage. In spite of Europe's early engagement with a welfare agenda, there are some sources of disagreement between European and North American zoos. For example, North American institutions regard euthanasia as an unsatisfactory technique for managing a population surplus, and recognize that the public will only support them if they are totally committed to solutions that also protect individual animal welfare. A recent news article in the *New York Times* thoroughly explored this divisive issue (Kaufman 2012). Sometimes our institutions must make gut-wrenching decisions, but careful, effective planning will tend to reduce or eliminate transactions that compromise welfare. In summarizing the recommendations from the *Ethics on the Ark* (1995) working groups in Atlanta, one assertion illustrates the prevailing position of North American zoos:

> Management euthanasia should be the last choice among the possibilities . . . If animals are euthanized because they are genetic surplus, there will never be resolution between zoos and animal protection societies. (p. 322)

9.6 Comparative Quality of Life

Whitham and Wielebnowski (2009) discussed applications of the Quality of Life (QOL) framework to focus on the life of the individual animal. QOL offers a holistic approach and accounts for the unique perspectives, preferences, and needs of individuals. In this context it may be possible to utilize animal-centered measures to examine subjective feelings. For many investigators, animal welfare is all about the feelings of individuals (Dawkins 1990, McMillan 2000). One way to assess feelings is to offer a close human caretaker as a proxy or voice for the animal. Qualitative assessments can generate testable hypotheses. Thus the experienced keeper, behavioral scientist, or veterinarian can detect subtle changes in psychological well-being that may not be detectable by an outside observer. King and Landau (2003) studied "happiness" in chimpanzees based on the human trait of "subjective well-being" (SWB). Despite variability in SWB within the sample, purposive upgrades in housing and environmental enrichment were effective regardless of the animal's unique personality. Similarly, chimpanzees confined in bare, unstimulating, or stressful surroundings clearly suffered from lowered SWB. Research by King and Landau supported the conclusion that to some degree humans and chimpanzees share common feelings of happiness, and human observers are able to recognize those feelings in a closely related species. Qualitative research measures are becoming more common in the assessment of animal emotions and in evaluations of the effects of environmental interventions. If there is any group of animals where happiness is a valid construct, it is the great apes; chimpanzees, gorillas, and orangutans. If observers can reliably rate the factor, we need to figure out how to consistently measure, monitor, and facilitate happiness in apes and other species. Our close phylogenetic relationship to apes is an advantage in reading their emotions but it is also a source of bias that may keep us from rendering objective decisions.

A new study by scientists at the University of Kent (Birkett and Newton-Fisher 2011) revealed signs of compromised mental health in a large sample of chimpanzees living in the collections of six accredited zoos in North America and the United Kingdom. The investigators suggested that captivity itself was the prime cause of abnormal behavior and speculated that chimpanzees may not be suitable for life in a zoo. This conclusion has been challenged by Ross and Bloomsmith (2011). While every zoo animal can benefit from enrichment and habitat improvements, some species such as chimpanzees surely require more substantial investment than others. This may be why so few zoos continue to exhibit chimpanzees. A successful exhibition of any of the great apes requires a highly complex and durable naturalistic environment, beyond the financial capacity of many zoos. Given their close genetic relationship to humankind, it would not be surprising if zoos were prohibited by governments from exhibiting chimpanzees in the future. In the meantime, captive chimpanzees deserve our best efforts to provide a superior quality of life.

The idea that some captive chimpanzees may experience a type of "mental illness" is not new, but it is a difficult and delicate subject. The chimpanzee's formidable intellectual capacity makes a comparison to human mental illness a reasonable possibility. Unique research by Lilienfeld and his colleagues (1999) examined the phenomenon of "psychopathic personality" to see if this construct applied to chimpanzees. The investigators, a team of clinical psychologists and experts in animal behavior, developed an instrument to measure psychopathy, the Chimpanzee Psychopathy Measure (CPM) and asked raters to use it to study 34 chimpanzees. The results provided evidence for the reliability and construct validity of the CPM, but there are many complications that make the interpretation preliminary at best. It is possible that the CPM and other personality instruments may be useful in studying their propensity for aggression, dominance, and even Machiavellian behavior that has been attributed to them. The fact that some chimpanzees have inexplicably and viciously attacked their caretakers and their peers is a troubling indication that there may be a connection between personality and violent behavior. Clearly, the complex mind and emotions of chimpanzees must be taken into account as we attempt to provide for their needs in the zoo. Their innate capacity for aggression in the wild must also be factored into our management equation. We acknowledge that many chimpanzees in modern zoos arrived at our institutions after living in other captive settings where they developed idiosyncratic, abnormal, and maladaptive behaviors that are resistant to change by enrichment or any other means. Although the rehabilitation and re-socialization of chimpanzees is challenging, the U.S. National Institutes of Health have repeatedly funded facilities to provide for mental health interventions. Opened in 1981, the chimpanzee facilities at the University of Texas System Cancer Center were designed to contribute to the psychological well-being of animals removed from biomedical research settings (Bloomsmith et al. 1988). Due to federal legislation by the United States government, a sanctuary for chimpanzees retired from biomedical service was established in Keithville, Louisiana. Chimp Haven was opened in 2003 and currently provides a home for more than 100 animals. The medical and behavioral scientists that operate Chimp Haven are committed to improving the lives of these animals regardless of their limitations, history, or handicaps. The sanctuary provides managed social opportunities, enrichment of all types, and large, naturalistic habitats. A helpful partner to zoos, Chimp Haven contributes valuable knowledge and expertise to the advancement of welfare for all captive chimpanzees. There is now a critical mass of great ape expertise in zoos and specialized primate facilities, and good reason to communicate and cooperate to the benefit of apes in all settings.

We have reason to be concerned about the effects of captivity on other megafauna. For example, the findings of Clubb et al. (2008) determined that captive elephants experienced a lifespan one-third as long as their wild counterparts in Africa and Asia. The authors reported that African elephants in Amboseli National Park in Kenya lived an average of 56 years whereas African elephants in zoos had life spans of approximately 17 years. Infant mortality was very high in zoos according to this study. In contrast, data from Wiese and Willis (2004) found essentially no difference in the longevity of Asian elephants in the zoo and the wild. Their research used life-table analysis to estimate median survivorship, finding that the median life expectancy for female Asian elephants is 35.9 years in North America and 41.9 years in Europe. With fewer data for African elephants, Wiese and Willis calculated life-expectancy estimates for North America at 33 years. A different statistical analysis may account for the difference in findings, but further research will be necessary to determine if there is a trend that should concern us. Wiese and Willis acknowledged that changes in husbandry practices have improved life-expectancy for zoo elephants, so the movement to superior practices and standards and larger, naturalistic facilities will likely establish that elephants can experience a natural lifespan in a zoo. For example, the San Diego Zoo manages a group of seven Asian elephants all in their late 40s and 50s.

Clubb's data clearly demonstrate that life in the wild is risky, as 142 of the 1,089 Amboseli elephants were killed by ritual tribal spearing, gunshot wounds, or other mishaps, demonstrating that survivorship is not the only valid indicator of quality of life. Prioritizing welfare and conservation will likely change the distribution of animals in zoos, and reduce collections to a size and scope that encourages improvements in facilities, husbandry, and management for the animals that can be exhibited well. While we are planning and building superior exhibits, we should retire exhibits that compromise animal welfare. Resource-intensive species such as elephants and chimpanzees demand nothing less than superior facilities.

In an earlier contribution to the literature of animal welfare science, Clubb and Mason (2003) demonstrated that a particular lifestyle in the wild confers vulnerability to the effects of captivity as measured by stereotypies and infant mortality. Among carnivores, they found that naturally wide-ranging species (e.g. arctic fox, polar bears, lions) exhibited the most evidence of stress, frustration, and/or psychological dysfunction in captivity. In their view, zoos should dramatically improve captive environments or phase out those species that don't prosper in captivity.

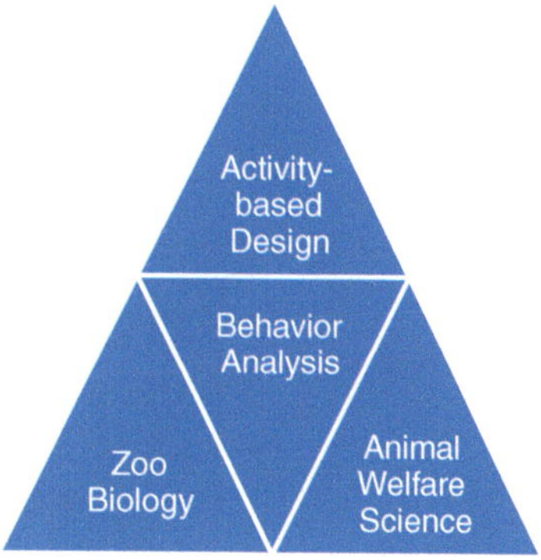

Fig. 9.5 Framework for institutional zoo animal welfare

There is clear evidence that zoos are responding to this issue by doing both. Although we don't always agree with our critics, it is essential that objective scientific work continue from experts outside the zoo profession. The scientific talent within zoos is sufficient to debate outside critics, but it is unhealthy to leave research only to those who have a vested interest in the reputation of the institution. Constructive criticism makes zoos stronger, smarter, and better, although there is always a risk that a friendly critic will turn out to be a troublesome adversary.

9.7 Going the Distance

The achievement of optimal zoo animal welfare is no longer a far distant goal. We have already achieved it for many species. We are confident that the science of animal welfare will continue to inform, shape, and guide our progress. The empirical zoo will always be the ideal template for innovation and change (Fig. 9.5), and we should expect to deliver future zoos and aquariums that are scarcely imaginable in our time. Scientific programs strengthen and protect the foundation of ethical arks, but they may not guarantee full protection. Wuichet and Norton (1995) suggested that zoos need a fifth directive (in addition to conservation, education, science, and recreation) to function as a positive, proactive goal to advance individual animal welfare. In their view, therefore, welfare should be regarded as co-equal to other zoo operational priorities. Management standards and practices, and exhibit reforms must evolve in bold new directions for the benefit of our zoo populations, and the millions of visitors whose support and advocacy are now an urgent necessity. By proclaiming and celebrating a renewed focus on animal welfare, zoos and aquariums can strengthen their reputations and regain public confidence. In the spirit that this book was written, we invite our readers to join with their local zoos and aquariums to support their commitment to health and welfare locally and globally. The enthusiasm of zoo professionals and their communities will determine whether zoo animal welfare will be a passing trend or a permanent and unyielding commitment.

References

Acampora RR (2007) Call for Papers. Farewell to Noah: Transforming animal encounters in the twilight of the zoo. Society Animals 15:213

Adams KM (2007) Refinement in the literature: searching for environmental enrichment. AATEX 14, Special Issue, 302–312

Agoramoorthy G (2004) Ethics and welfare in Southeast Asian zoos. J Appl Anim Welf Sci 7 (3):189–195

Agoramoorthy (2010) Setting standards for evaluation of captive facilities, Southeast Asia. In: Kleiman et al. (Eds) Wild mammals in captivity. University of Chicago Press, Chicago/ London, p 28–31

Altman I (1975) The environment and social behavior. Belmont, CA, Wadsworth

Altmann JE (2008) Research at a local zoo. Teachpsych.org/e-books, 151–155

Altmann JE, Gross KL, Lowry SR (2005) Nutritional and behavioral effects of gorge and fast feeding in captive lions. J Appl Anim Welf Sci 8:47–57

American Psychological Association Online. Journal of Comparative Psychology, journal description

Anderson US, Maple TL, Bloomsmith MA (2010) Factors facilitating research: a survey of zoo professionals. Zoo Biol 29(6):663–675

Appleby MC, Mench JA, Olsson IAS, Hughes BO (eds) (2011) Animal Welfare, 2nd edn. CAB International, UK/USA

Asa CS, Traylor-Holzer K, Lacy RC (2011) Can conservation-breeding programmes be improved by incorporating mate choice? Int Zoo Yearbook 45:203–212

Asher L, Collins LM, Ortiz-Pelaez A, Drewe JA, Nicol CJ, Pfeiffer DU (2009) Recent advances in the analysis of behavioural organization and interpretation as indicators of animal welfare. J Royal Soc Inter 1–17

Baker K (1997) Straw and forage material ameliorate abnormal behavior in adult chimpanzees. Zoo Biol 16:225–236

Barber J (2009) Programmatic approaches to assessing and improving animal welfare in zoos and aquariums. Zoo Biol 28:519–530

Barber JCE, Mellen J (2008) Assessing animal welfare in zoos and aquariums: is it possible? In: Bettinger T, Bielitzki J (eds) The well-being of Animals in Zoo and Aquarium sponsored research: putting best practices forward. Scientists Center for Animal Welfare, Greenbelt, MD, pp 39–52

Barnett JL, Hemsworth PH (2009) Welfare monitoring schemes: Using research to safeguard welfare of animals on the farm. J Appl Anim Welf Sci 12:114–131

Bashaw MJ, Tarou LR, Maki TS, Maple TL (2001) A survey assessment of variables related to stereotypy in captive giraffe and okapi. Appl Animal Behav Sci 73:235–47

T.L. Maple and B.M. Perdue, *Zoo Animal Welfare*, Animal Welfare,
DOI 10.1007/978-3-642-35955-2, © Springer-Verlag Berlin Heidelberg 2013

Bashaw MJ, Bloomsmith MA, Maple TL (2003) To hunt or not hunt? A feeding enrichment experiment with captive large felids. Zoo Biol 22:189–198

Bates LA, Saylalel KN, Njiraini NW, Moss CJ, Poole JH, Byrne RW (2007) Elephants classify human ethnic groups by odor and garment color. Curr Biol 17:1–5

Baumeister AA, Forehand R (1973) Stereotyped acts. Int Rev Res Ment Retard 6:55–96

Beck BB (1980) Animal tool behavior. Garland STPM Publ, New York

Beck BB, Power M (1988) Correlates of sexual and maternal competence in captive gorillas. Zoo Biol 7:339–350

Beck BB, Stoinski TS, Hutchins M, Maple TL, Norton B, Rowan A, Stevens EF, Arluke A (eds) (2001) Great Apes and Humans: The ethics of coexistence. Smithsonian Institution Press, Washington and London

Bekoff M (2010) First do no harm. New Scientist 24–25

Berkson G, Mason WA (1964) Stereotyped behaviors of chimpanzees: relation to general arousal and alternative activities. Percept Mot Skills 19:635–652

Benirschke K (ed) (1986) Primates: the road to self-sustaining populations. Springer/Verlag, New York

Beston H (1928) The outermost house. Doubleday and Doran Co, New York

Bettinger T, Wallis J, Carter T (1994) Spatial selection in captive adult female chimpanzees. Zoo Biol 13:167–176

Bingham HC (1932) Gorillas in a natural habitat, vol 426. Carnegie Institute, Washington, DC, pp 1–66

Birkett LP, Newton-Fisher NE (2011) How abnormal is the behavior of captive, zoo-living chimpanzees? PLoS ONE 6(6):e20101 OI, published online

Blaney E, Wells D (2004) The influence of a camouflage net barrier on the behavior, welfare, and public perceptions of zoo-housed gorillas. Animal Welfare 13:111–118

Bloomsmith MA (1989) Feeding enrichment for captive great apes. In: Segal EF (ed) Housing, care, and psychological well-being of captive and laboratory primates. Noyes Publications, Park Ridge, NJ, pp 336–356

Bloomsmith MA (1995) Evolving a behavioral management program. Shape Enrich 4(1):7–10

Bloomsmith MA, Lambeth SP (2000) Videotapes as enrichment for captive chimpanzees. Zoo Biol 6:541–551

Bloomsmith MA, Alford PL, Maple TL (1988) Successful feeding enrichment for captive chimpanzees. Am J Primatol 16(2):155–164

Bloomsmith MA, Keeling ME, Lambeth SP (1990) Videotapes: Environmental enrichment for singly housed chimpanzees. Lab Anim 19:42–46

Bloomsmith MA, Brent LA, Schapiro SJ (1991) Guidelines for developing and managing an environmental enrichment program for nonhuman primates. Lab Anim Sci 41:372–378

Bloomsmith MA, Jones ML, Snyder RJ, Singer RA, Gardner WA, Liu SC, Maple TL (2003) Positive reinforcement training to elicit voluntary movement of two giant pandas throughout their enclosure. Zoo Biol 22:323–334

Bloomsmith MA, Marr MJ, Maple TL (2007) Addressing nonhuman primate behavioral problems through the use of operant conditioning: is the human treatment approach a useful model? J Appl Animal Behav Sci 102:205–222

Blowers TE, Waterman JM, Kuhar CW, Bettinger TL (2012) Female nile hippopotamus space use in a naturalistic exhibit. Zoo Biol 31:129–136

Blum D (2002) Love at Goon Park: Harry Harlow and the science of affection. Perseus Publishing, Cambridge, MA

Bolwig N, Hill DH, Philpott M (1965) Hand-rearing an African elephant *Loxodonta africana*. Int Zoo Yearbook 5:152–154

Bond MBR, Block JA (1982) Growth and development of twin orangutans *Pongo pygmaeus*. Int Zoo Yearbook 22:256–261

Bracke MBM, Hopster H (2006) Assessing the importance of natural behavior for animal welfare. J Agr Environ Ethics 19:77–89

Bradshaw GA (2009) Inside looking out: neuroethological compromise effects in elephants in captivity. In: Forthman et al. (eds) The elephant in the room, pp 55–68
Bradshaw GA, Sapolsky RM (2006) Mirror, mirror. American Scientist 487–489
Braithwaite V (2010) Do fish feel pain? Oxford University Press, Oxford, UK
Brambell FWR (1965) Report of the technical committee to enquire into the welfare of animals kept under intensive livestock husbandry systems, the Brambell Report. Her Majesty's Stationary Office, London
Bramblett C (1989) Mental wellbeing in anthropoids. In: Segal EF (ed) Housing, Care and psychological wellbeing of captive and laboratory primates. Noyes Publications, Park Ridge, NJ, pp 1–12
Brockett R, Stoinski T, Black J, Markowitz T, Maple TL (1999) Nocturnal behavior in a group of unchained African elephants. Zoo Biol 18(2):101–109
Bronwyn B, Ford JC (1991) Environmental enrichment in zoos: Melbourne Zoo's naturalistic approach. Thylacinus 16(1):12–17
Brown SG, Dunlap W, Maple TL (1982) Notes on the water-contact behavior of captive lowland gorillas. Zoo Biol 1(2):243–249
Brown BB, Werner CM, Amburgey JW, Szalay C (2007) Walkable route perceptions and physical features: converging evidence for en route walking experiences. Environ Behav 39(1):34–61
Brydges NM, Leach M, Nicol K, Wright R, Bateson M (2011) Environmental enrichment induces optimistic cognitive bias in rats. Anim Behav 81(1):169–175
Burgener N, Gusset M, Schmid H (2008) Frustrated appetitive foraging behavior, stereotypic pacing, and fecal glucocorticoid levels in snow leopards (Uncia uncia) in the Zurich Zoo. J Appl Anim Welf Sci 11:74–83
Burghardt G, Ward B, Roscoe R (1996) Problem of reptile play: environmental enrichment and play behavior in captive Nile soft-shelled turtle, Trionyx triunguis. Zoo Biol 15:223–38
Burks K, Mellen JD, Miller GW, Lenhardt J, Weiss A, Figueredo AJ, Maple TL (2004) Zoo Biology 23(2):109–126
Burman OHP, Parker RMA, Paul ES, Mendl MT (2009) Anxiety-induced cognitive bias in non-human animals. Physiol Behav 98:345–350
Calhoun JC (1962) Population density and social pathology. Scientific American, CCVI, 139–148
Canino W, Powell D (2010) Formal behavioral evaluation of enrichment programs on a zookeeper's schedule: a case study with a polar bear (Ursus maritimus) at the Bronx Zoo. Zoo Biol 29:503–508
Caras R (1995) A view from the ASPCA. In: Norton BG et al (eds) Ethics on the ark. Smithsonian Institution Press, Washington/London, pp 297–299
Carlstead K (2009) A comparative approach to the study of keeper-animal relationships in the zoo. Zoo Biol 28:589–608
Carlstead K, Brown JL (2005) Relationships between patterns of fecal corticoid excretion and behavior, reproduction and environmental factors in captive black (Diceros bicornis) and white (Ceratotherium simum) rhinoceros. Zoo Biol 24:215–232
Carlstead K, Fraser J, Kleiman DG (1999) Black rhinoceros (Diceros bicornis) in U.S. zoos: II. Behavior, breeding success and mortality in relation to housing facilities. Zoo Biol 18:35–52
Carpenter CR (1937) An observational study of two captive mountain gorillas (Gorilla beringei). Hum Biol 9:175–196
Carpenter CR (1942) Sexual behavior of free-ranging rhesus monkeys. J Comp Psychol 33:113–142
Carpenter CR (1964) Naturalistic behavior of nonhuman primates. Pennsylvania State University Press, State College, PA
Chamove A, Moodie E (1990) Brief threatening events beneficial for captive tamarins? Zoo Biol 9:275–286
Chamove AS, Anderson JR (1989) Examining environmental enrichment. In: Segal EF (ed) Housing, Care, and psychological wellbeing of captive and laboratory primates. Noyes Publications, Park Ridge, NJ, pp 183–202

Chamove AS, Hosey GR, Schaetzel P (1988) Visitors excite primates in zoos. Zoo Biol 7:359–369
Chang T, Forthman DL, Maple TL (1999) Comparisons of captive mandrills in traditional versus ecologically-relevant exhibits. Zoo Biol 18(3):163–176
Clarke AS, Juno CJ, Maple TL (1982) Behavioral effects of a change in the physical environment: a pilot study of captive chimpanzees. Zoo Biol 1:65–70
Clay AW (2011) Attachment and early rearing: longitudinal effects in Chimpanzees. Doctoral Dissertation, Georgia Institute of Technology, Atlanta, GA
Clay AW, Perdue BM, Maple TL, Bloomsmith MA (2007) Transfer of food preferences to photographs in captive western lowland gorillas (Gorilla gorilla gorilla). Am J Primatol 71 (Suppl):1–27
Clay AW, Perdue B, Maple TL, Bloomsmith MA (2009) Transfer of food preferences to photographs in captive western lowland gorillas (Gorilla gorilla gorilla). Abstract from the American Society of Primatologists Annual Meeting
Clay AW, Perdue BM, Gaalema DE, Dolins F, Bloomsmith MA (2011) The use of technology to enhance zoological parks. Zoo Biol 30:487–497
Clubb R, Mason G (2003) Captivity effects on wide-ranging carnivores. Nature 425:473–474
Clubb R, Rowcliffe M, Lee P, Mar KY, Moss C, Mason GJ (2008) Compromised survivorship in zoo elephants. Science 322(5908):1649
Coe JC (1985) Design and perception: making the zoo experience real. Zoo Biol 4:197–208
Coe JC (1996) Future fusion: the twenty-first century zoo. In: Nichols M (ed) Keepers of the wild. Thomasson-Grant & LIckie, Charlottesville, VA, pp 105–115
Coe JC (2003) Steering the ark toward Eden: design for animal well-being. Animal welfare forum: The welfare of zoo animals. JAVMA 223(7):977–980
Coe JC, Scott D, Lukas KE (2009) Facility design for bachelor gorilla groups. Zoo Biol 28:144–162
Colahan H, Breder C (2003) Primate training at Disney's Animal Kingdom. J Appl Anim Welf Sci 6(3):235–246
Collins J (2001) Good to great. Harper, New York
Collins J Good to Great and the Social Sector. Harper, New York
Conway WG (1995) Zoo conservation and ethical paradoxes. In: Norton et al. (eds) Ethics on the ark. Smithsonian, London and Washington, pp 1–9
Conway WG (1999) Congo: a zoo experiment in participatory conservation. Proceedings of the annual meeting of the American Zoo and Aquarium Association, pp 101–104
Conway WG (2011) Buying time for wild animals with zoos. Zoo Biol 30:1–8
Coppernoll C (2011) Oklahoma City elephant exhibit debuts as largest in the country. www.NewsOK.com
Cornick LA, Markowitz H (2002) Diurnal vocal patterns of the black howler monkey at Lamanai, Belize. J Mammal 83(1):159–166
Cottle L, Tamir D, Hyseni M, Buhler D, Lindemann-Mathies P (2010) Feeding live prey to zoo animals: response of zoo visitors in Switzerland. Zoo Biol 29:344–350
Cousins D (1979) Mortality factors in captive gorillas. Int Zoo News 30:5–17
Crockett CM (1996) Data collection in the zoo setting, emphasizing behavior. In: Kleiman et al. (eds) Wild mammals in captivity. University of Chicago Press, London and Chicago, pp 545–565
Cummings D, Brown JL, Rodden MD, Songsasen N (2007) Behavioral and physiologic responses to environmental enrichment in the maned wolf (Chrysocyon brachyurus). Zoo Biol 26:331–343
Cuppy W (1931) How to tell your friends from the apes. Liveright, New York
Dallaire JA, Field N, Mason GJ (2012) Activity and enrichment use in disabled Asiatic black bears rescued from bile farms. Animal Welfare 21:167–176
Davenport RK (1979) Some behavioral disturbances of great apes in captivity. In: Hamburg D, McKown ER (eds) The great apes. Benjamin/Cummings, Menlo Park, CA, pp 341–356

Davenport RK, Rogers CM, Rumbaugh DM (1973) Long-term cognitive deficits in chimpanzees associated with early impoverished rearing. Dev Psychol 9(3):343–347
Davey G (2007) Visitors' effects on the welfare of animals in the zoo: a review. J Appl Anim Welf Sci 10(2):169–183
Dawkins MS (1990) From an animal's point of view; motivation, fitness, and animal welfare. Behav Brain Sci 13:1–61
Dawkins MS (2008) The science of animal suffering. Ethology 114:937–945
De Azevedo CS, Cipreste CF, Young RJ (2007) Environmental enrichment: a GAP analysis. Appl Animal Behav Sci 102:339–343
De Boer L (ed) (1979) The orang utan: its conservation and biology. Dr. W. Junk, The Hague
De Waal FBM (1978) Exploitative and familiarity dependent support strategies in a colony of semi-free ranging chimpanzees. Behaviour 66:268–31
Dembiec DP, Snider RJ, Zanella AJ (2004) The effects of transport stress on tiger physiology and behavior. Zoo Biol 23:335–346
Desmond T, Laule G (1991) Protected contact elephant training. Proceedings of AAZPA Annual Conference, Wheeling, WVA, pp. 606–613
Desiderio F (2000) Raising the bars: the transformation of Atlanta's Zoo, 1889–2000. Atlanta History Xliii(4):1–64
Dewsbury DA (1972) Patterns of copulatory behavior in male mammals. Q Rev Biol 47:1–33
Dewsbury DA, Pierce JD Jr (1989) Copulatory behavior of primates as viewed in broad mammalian perspective. Am J Primatol 17:51–72
Dierenfeld ES, Kilbourn A, Karesh W, Bosi E, Andau M, Alsisto S (2006) Intake, utilization, and composition of browses consumed by the Sumatran rhinoceros in captivity in Sabah, Malaysia. Zoo Biol 25(5):417–431
Dybas CL (2007) Out of Africa: A tale of gorillas, heart disease and a swamp plant. Bioscience 57(5):392–397
Estep DQ, Dewsbury DA (1996) Mammalian reproductive behaviour. In: Kleiman DG et al (eds) Wild mammals in captivity. University Chicago Press, Chicago, IL, pp 379–389
Edwards TL, Poling A (2011) Animal research in the Journal of Applied Behavior Analysis. J Appl Behav Anal 44(2):409–412
Ellenberger HF (1960) Zoological garden and mental hospital. Can Psychiatr Assoc J 5:136–149
Erwin J (1979) Aggression in captive macaques. In: Erwin J et al (eds) Captivity and behavior. Van Nostrand Reinhold Co, New York, pp 139–171
Erwin J, Maple TL, Mitchell G (eds) (1979) Captivity and behavior: Primates in breeding colonies, laboratories and zoos. Van Nostrand Reinhold Co, New York
Fabregas MC, Guillen-Salazar F, Garces-Narro C (2012) Do naturalistic enclosures provide suitable environments for zoo animals? Zoo Biol 31:362–373
Farm Animal Welfare Council (1979) Press Release. http://www.fawc.org.uk/pdf/fivefreedoms1979.pdf Accessed 11 July 2012
Fernandez LT, Bashaw MJ, Sartor RI, Bouwens NR, Maki TS (2008) Tongue twisters: feeding enrichment to reduce oral stereotypy in giraffe. Zoo Biol 27:200–212
Fernandez EJ, Tamborski MA, Pickens SR, Timberlake W (2009) Animal-visitor interactions in the modern zoo: conflicts and interventions. Appl Animal Behav Sci 120:1–8
Fernie AC, Tribe A, Murray PJ, Lisle A, Phillips CJC (2012) A survey of the attitudes of stakeholders in the zoo industry towards the husbandry requirements of captive great apes. Animal Welfare 21:233–245
Fiby M (2005) Zurich Zoo Masoala Rainforest. ZooLex online
Fiby M, Berthier J (2007) Environment enrichment incorporated into masterplanning. ZooLex online
Fiby M, Petzold D (2005) Zoo design: replication – innovation – evolution. Information management by www.zoolex.org. In: Plowman and Tonge (eds) Innovation or Replication. Whitley Wildlife Conservation Trust, Paignton/Devon/UK, pp 71–74

Fielder FE, Wheeler WA (1985) A survey of visitors in the Woodland Park Zoological Gardens. The Zoological Foundation of Woodland Park, Seattle, WA

Finlay T, James LR, Maple TL (1988) People's perceptions of animals: the influence of zoo environments. Environ Behav 20(4):508–528

Forthman DL, Kane LF, Hancocks D, Waldau PF (eds) (2009) An elephant in the room: the science and well-being of elephants in captivity. Center for Animals and Public Policy, Tufts University, North Grafton, MA

Forthman DL, Ogden JJ (1992) The role of applied behavior analysis in zoo management: today and tomorrow. J Appl Behav Anal 25(3):647–652

Forthman DL (1998) Toward optimal care for confined ungulates. In Second Nature, pp 236–261

Forthman DL, Elder SD, Bakeman R, Kurkowski T, Noble CC, Winslow SW (1992) Effects of feeding enrichment on three species of captive bears. Zoo Biol 11:187–196

Forthman DL, McManamon R, Levi UL, Bruner GY (1995) Interdisciplinary issues in the design of mammal exhibits. In: Gibbons EF Jr, Demarest J, Durrant B (eds) Captive conservation of endangered species. State University Press, Albany, NY, pp 377–399

Forthman-Quick DL (1984) An integrative approach to environmental engineering in zoos. Zoo Biol 3:29–43

Forthman-Quick DL, Pappas TC (1986) Enclosure utilization, activity budgets, and social behavior of captive chamois (Rupicapra rupicapra) during the rut. Zoo Biol 5:281–292

Fox MW (ed) (1968) Abnormal behavior of animals. W.B. Saunders, Philadelphia, PA

Fraser D (2009) Assessing animal welfare: Different philosophies, different scientific approaches. Zoo Biol 28:507–518

Frediani K (2009) Exploring the potential for plant based enrichment. Proceedings international conference of environmental enrichment. Published in ZooLex online

Fritz J, Howell SM (1993) Psychological wellness for captive chimpanzees. Hum Innov Altern 7:426–434

Fuller G, Sadowski L, Cassella C, Lukas KE (2010) Examining deep litter as environmental enrichment for a family group of Wolf's guenons. Zoo Biol 29(5):626–632

Gaalema DE, Perdue BM, Kelling AS (2011) Food preference, keeper ratings, and reinforcer effectiveness in exotic animals: The value of systematic testing. J Appl Anim Welf Sci 14(1):33–41

Galindo F, Newberry RC, Mendl M (2011). Social conditions. In: Appleby et al. (eds) Animal welfare, 2nd edn. CAB International, UK and USA, pp 228–245

Geller ES (1985) Corporate safety belt programs. Virginian Polytechnic Institute and State University, Blacksburg, VA

Gladwell M (2008) Outliers. Little Brown, New York

Glatson AR, Geilvoet-Soeteman E, Hora-Pacek E, van Hooff JARAM (1984) The influence of the zoo environment and social behavior of groups of cotton-topped tamarins, Saguinus oedipus oedipus. Zoo Biol 3(3):241–254

Gold KC, Maple TL (1994) Personality assessment in the gorilla and its utility as a management tool. Zoo Biol 13:509–522

Goldman JG (2012) A psychologist goes to the zoo: an interview with Terry Maple. Scientific American online

Gonyou HW (1994) Why the study of animal behavior is associated with the animal welfare issue. J Anim Sci 72:2171–2177

Goodall J (2001) Problems faced by wild and captive chimpanzees: finding solutions. In: Beck BB, Stoinski TS, Hutchins M, Maple TL, Norton B, Rowan A, Stevens EF, Arluke A (eds) Great apes & humans: The ethics of coexistence. Smithsonian Institution Press, Washington and London, p xiii-2

Goymann W, Mostl E, Van't Hof T, East ML, Hofer H (1999) Noninvasive fecal monitoring of glucocorticoids in spotted hyenas, Crocuta crocuta. Gen Comp Endocrinol 114:340–348

Grant AM, Schwartz B (2011) Too much of a good thing: the challenge and opportunity of the inverted U. Perspect Psychol Sci 6(1):61–76

Grand AP, Kuhar WC, Leighty KA, Bettinger TL, Laudenslager ML (2012) Appl Anim Behav Sci 142(1–2):69–75
Greene M (1987) No rms, jungle vu. Atlantic Monthly 260(6):62–78
Griede T (1992) 200 examples of environmental enrichment for zoo animals. National Foundation for Research in Zoological Gardens, The Amsterdam
Gruen L (2011) Ethics and animals. UK, Cambridge University Press, Cambridge
Gupta BK, Chakraborty B (2005) The role of zoos in the rehabilitation of animals in the circus. J Appl Anim Welf Sci 8(4):285–294
Hafez ESE (ed) (1962) The behaviour of domestic animals. Balliere-Tindall, London
Hafez ESE (ed) (1969) The behaviour of domestic animals, 2nd edn. Balliere-Tindall, London
Hancocks D (1995) Lions and tigers and bears, Oh No! In Norton et al. (eds) Ethics on the ark. Smithsonian Press, Washington, DC, pp 31–37
Hancocks D (2007) The future and ethics of zoos. Presentation to Museums Australia National Conference, Canberra, Australia, May, pp 1–8
Hancocks D (2001) A different nature: The paradoxical world of zoos and their uncertain future. University of California Press, Berkeley, CA
Hancocks D (2009) What will new zoo environments look like? In: Forthman DL et al (eds) An elephant in the room. Center for Animals in Public Policy, Tufts University, North Grafton, MA
Hancocks D (2012) The future of zoos. Unpublished ms
Harcourt AH (1987) Behaviour of wild gorillas and their management in captivity. Int Zoo Yearbook 26:248–255
Harcout AH, Stewart KJ (1978) Coprophagy by wild mountain gorillas. East African Wildlife Soc J 16:223–225
Hardin CJ (1976) Hand-rearing a giant anteater Myrmecophaga tridactyla at Toledo Zoo. Int Zoo Yearbook 16:199–200
Hare VJ, Ripsky D, Battershill R, Bacon K, Hawk K, Swaisgood RR (2003) Giant panda enrichment: meeting everyone's needs. Zoo Biol 22(4):401–416
Harlow HF (1971) Learning to love. Aldine, Chicago
Harlow HF, Harlow MK (1965) The affectional system. In: Schrier AM, Harlow HF, Stollnitz F (eds) Behaviour of nonhuman primates 2. Academic Press, New York
Hayes MP, Jennings MR, Mellen JD (1998) Environmental enrichment for amphibians and reptiles. In: Shepherdson et al. (eds) Second nature. Smithsonian Institution Press, Washington and London, pp 205–235
Hediger H (1955) Studies of the psychology and behaviour of animals in zoos and circuses. Butterworths, London
Hediger H (1964) Wild animals in captivity. Butterworths, London
Hediger H (1970) Man and animal in the zoo. Delacorte Press, New York
Herman LM (1987) Receptive competencies of language trained animals. In: Rosenblatt JS, Beer C, Busnel MC, Slater PBJ (eds) Advances in the study of behavior, vol 17. Academic, New York, pp 1–60
Highfill L, Hanbury D, Kristiansen R, Kuczaj S, Watson S (2010) Rating vs. coding in animal personality research. Zoo Biol 29:509–516
Hinde RA, Spencer-Booth Y (1967) The behaviour of socially living rhesus monkeys in their first two and a half years. Anim Behav 15:169–196
Hoage RJ (1996) Epilogue. In: Hoage RJ, Deiss WA (eds) New worlds, new animals: From menagerie to zoological park in the nineteenth century. Johns Hopkins University Press, Baltimore and London, pp 136–137
Hoff MP, Tarou LR, Horton C, Mayo L, Maple TL (2005) Notes on the introduction of an 11 week-old infant western lowland gorilla to a non-lactating surrogate mother at Zoo Atlanta. Int Zoo Yearbook 39:191–198

Holden MD, Gregory J, Watkins V, Radford L (2006) Operant-conditioning programme for white rhinoceros, black rhinoceros, and Indian or greater one-horned Asian rhinoceros at Whipsnade Wild Animal Park, Dunstable, UK. Int Zoo Yearbook 40:144–149

Holzapfel M (1938) Uber bewegungsstereotypien bei gehaltenen saugern. Z Tierpsychologie 2:46–72

Horikoshi-Beckett C, Schulte BA (2006) Activity patterns and spatial use of facility by a group of captive female manatees (Trichechus manatus latirostris). Zoo Biol 25:285–301

Hosey G (1997) Behavioral research in zoos: academic perspectives. Appl Animal Behav Sci 51:199–207

Hosey G (2005) How does the zoo environment affect the behavior of captive primates? Appl Animal Behav Sci 90:107–129

Hosey G, Melfi V (2012) Human-animal bonds between zoo professionals and he animals in their care. Zoo Biol 31:13–26

Howard HE (1920) Territory in bird life. EP Dutton, New York

Hoy JM, Murray PJ, Tribe A (2010) Thirty years later: enrichment practices for captive mammals. Zoo Biol 29:303–316

Humane Society of the United States (2012a). Poll reveals Americans oppose keeping orcas in captivity for public display. Online 16 July 2012

Humane Society of the United States (2012b) The cruel trade in beluga whales needs to end. Online 15 October 2012

Hunt H, Ogden JJ (1991) Selected aspects of the nesting ecology of American alligators in the Okefenokee Swamp. J Herpetol 28(4):448–453

Hutchins M (2006) Death at the zoo: the media, science, and reality. Zoo Biol 25:101–115

Hunter LTB, White P, Henschel P, Frank L, Burton C, Loveridge A, Balme G, Breitenmoser C, Breitenmoser U (2012) Walking with lions: why there is no role for captive-origin lions Panthera Leo in species restoration. Oryx online, 30 September 2012, pp 1–6

Hutt C, Hutt SJ (1965) Effects of environmental complexity on stereotyped behaviours of children. Anim Behav 13:1–4

Iosso G, Soulsbury CD, Harris S (2009) Are wild animals suited to a travelling circus life? Animal Welfare 18:129–140

Jackson DW (1996) Horticultural philosophies in zoo exhibit design. In: Kleiman DG et al (eds) Wild mammals in captivity. University of Chicago Press, Chicago and London, pp 175–179

Jamieson D (1985) Against zoos. In: Singer P (ed) In defense of animals. Blackwell, Oxford, pp 109–117

Jamieson D (1995) Zoos revisited. In: Norton BG et al (eds) Ethics on the ark. Smithsonian Institution Press, Washington/London, pp 52–66

Jenny S, Schmid H (2002) Effect of feeding boxes on the behavior of stereotyping Amur tigers in the Zurich Zoo. Zoo Biol 21:573–584

Jones GR (1982) Design principles for presentation of animals and nature. AAZPA annual conference proceedings. American Association of Zoological Parks and Aquariums, Wheeling, WV, pp 184–192

Jones GR, Coe JC, Paulson DR (1976) Woodland Park Zoo: Long range plan, development guidelines, and exhibit scenarios. Jones & Jones for Seattle Department of Parks and Recreation, Seattle, WA

Jordan MJR (2005) Dietary analysis for mammals and birds: a review of field techniques and animal-management applications. Int Zoo Yearbook 39:108–116

Kagan R, Veasey J (2010) Challenges of zoo animal welfare. In: Kleiman DG et al (eds) Wild animals in captivity, 2nd edn. University Chicago Press, Chicago, IL, pp 11–21

Kane LF, Forthman DL, Hancocks D (2009) Optimal conditions for captive elephants. In: Forthman DL et al (eds) An elephant in the room. Center for Animals and Public Policy, Tufts University, North Grafton, MA, pp 346–379

Kaufman IC, Rosenblum LA (1967) The reaction to separation in infant monkeys: anaclitic depression and conservation withdrawal. Psychosom Med 29:648–675

Kaufman L (2012) When babies don't fit plan, question for zoos is, now what? New York Times online, 2 August 2012, p 4

Kawata K (2008) Zoo animal feeding: A natural history viewpoint. Der Zoologische Garten 78:17–42

Keeling ME, Alford PL, Bloomsmith MA (1991) Decision analysis for developing programs of psychological well-being: a bias-for-action approach. In: Novak MA, Petto AJ (eds) Through the looking glass. American Psychological Association, Washington, DC, pp 57–65

Keeling LJ, Rushen J, Duncan IJH (2011) Understanding animal welfare. In: Appleby MC, Mench JA, Olsson IAS, Hughes BO (eds) Animal welfare, 2nd edn. CABI International, Oxfordshire, UK

Kelling AS (2008) An examination of salivary cortisol concentrations and behavior in three captive African elephants (Loxodonta africana) at Zoo Atlanta. PhD Dissertation, Georgia Institute of Technology, Atlanta, GA

Kelling AS, Snyder RL, Gardner W, Marr MJ, Bloomsmith MA, Maple TL (2006) Color vision in the giant panda. Learn Behav 34(2):154–161

Kelling AS, Allard SM, Kelling NJ, Sandhaus EA, Maple TL (2012) Lion, ungulate, and visitor reactions to playbacks of lion roars at Zoo Atlanta. J Appl Anim Welf Sci 15:1–16

Kellogg WN, Rice CE (1966) Visual discrimination and problem solving in a bottle-nosed dolphin. In: Norris KS (ed) Whales, dolphins, and porpoises. University of California Press, Berkeley, pp 731–754

King CE (1993) Environmental enrichment: is it for the birds? Zoo Biol 12:509–512

King JE, Landau VL (2003) Can chimpanzee happiness be estimated by human raters? J Res Personal 37:1–15

Kinzley C (2009) Applied behavioral research: a tool in captive elephant management. In: Forthman DL, Kane LF, Hancocks D, Waldau PF (eds) An elephant in the room. Tufts Center for Animals and Public Policy, North Grafton, MA

Kirchshofer R, Fradrich H, Podolczak D, Podolczak G (1967) An account of the physical and behavioral development of the hand-reared gorilla infant Gorilla g. gorilla born at Frankfurt Zoo. Int Zoo Yearbook 7:108–113

Kirkden RD, Pajor EA (2006) Using preference, motivation and aversion tests to ask scientific questions about animals' feelings. Appl Animal Behav Sci 100:29–47

Koch W (2007) Celebrating a life by recognizing realities. J Appl Anim Welf Sci 10(1):1–6

Koene P, Duncan JJH (2001) From environmental requirement to environmental enrichment: from animal suffering to animal pleasure. Proceedings of Fifth Annual Conference on Environmental Enrichment. Taronga Zoo, Sydney, Australia, p 36

Kohler W (1927) The intelligence of apes. Metheun, London

Kortlandt A (1960) Can lessons from the wild improve the lot of captive chimpanzees? Int Zoo Yearbook 2:76–80

Kreger MD, Hutchins M (2010) Ethics of keeping mammals in zoos and aquariums. In: Kleiman DG, Thompson KV, Baer CK (eds) Wild mammals in captivity. University of Chicago Press, Chicago and London

Kuczaj II SA, Lacinak CT, Turner TN (19xx) Environmental enrichment for marine mammals at Sea World. In: Shepherdson, et al. (eds) Second nature: Environmental enrichment for captive animals. Smithsonian, Washington and London, pp 314–328

Kuhar C (2006) In the deep end: pooling data and other statistical challenges of zoo and aquarium research. Zoo Biol 25:339–352

Kuhar CW, Stoinski TS, Lukas KE, Maple TL (2006) Gorilla behavior index revisited: age, housing, and behavior. Appl Animal Behav Sci 96:315–326

Kuhar C, Fuller GA, Dennis PM (2012) A survey of diabetes prevalence in zoo-housed primates. Zoo Biol 00:1–7

Kummer H (1968) Social organization of hamadryas baboons. Karger, Basel

Kutska D (2009) Variation in visitor perceptions of a polar bear enclosure based on the presence of natural vs. un-natural enrichment items. Zoo Biol 28:292–306

Latham N (2010) Brief introduction to welfare assessment. In: Hubrecht R, Kirkwood J(eds) The UFAW handbook on the care and management of laboratory and other research animals. Wiley, West Sussex

Laule G (1993) The use of behavioral management techniques to reduce or eliminate abnormal behavior. Animal Welfare Information Center Newsletter 4(4):1–2–8–11

Laule G, Desmond T (1998) Positive reinforcement as an enrichment strategy. In: Shepherdson DJ, Mellen JD, Hutchins M (eds) Second nature: environmental enrichment for captive animals. Smithsonian Institution Press, Washington/London, pp 302–313

Laws N, Ganswindt A, Heistermann M, Harris M, Harris S, Sherwin C (2007) A case study: Fecal corticosteroid and behavior as indicators of welfare during relocation of an Asian elephant. J Appl Anim Welf Sci 10:349–358

Lee PC (1986) Early social development among African elephant calves. Natl Geogr Res 2:388–401

Leighty KA, Soltis J, Savage A (2010) GPS assessment of the use of space and resources by African elephants (Loxodonta africana). Zoo Biol 29:210–220

Lemonick MD (2006). Who belongs in the zoo? Time Magazine, June 11, 3 pp.

Lewis KD, Shepherdson DJ, Owens TM, Keele M (2010) A survey of elephant husbandry and foot health in North American zoos. Zoo Biol 29:221–236

Lilienfeld SO, Gershon J, Duke M, Marino L, de Waal FBM (1999) A preliminary investigation of the construct of psychopathic personality (psychopathy) in chimpanzees. J Comp Psychol 113(4):365–375

Lindburg DG (1969) Rhesus monkeys: mating season mobility of adult males. Science 166:1176–1178

Lindburg DG (1988) Improving the feeding of captive felines through application of field data. Zoo Biol 7:211–218

Lindburg DG (1998) Enrichment of captive mammals through provisioning. In: Sherpherdson et al. (eds) Second nature: Environmental enrichment for captive animals. Smithsonian, Washington, DC, pp 262–276

Lindsay P et al (2012) Illegal hunting & the bushmeat trade in Savanna Africa: Drivers. Impacts & Solutions to Address the Problem, Panthera/WCS/ZSL/Traffic Report

Line SW, Markowitz H, Morgan KN, Strong S (1991) Effects of cage size and environmental enrichment on behavioral and physiological responses of rhesus macaques to the stress of daily events. In: Novak MA, Petto AJ (eds) Through the looking glass: Issues of psychological well-being in captive nonhuman primates. American Psychological Association, Washington, DC, pp 160–180

Linklater WL, MacDonald EA, Flamand JRB, Czekala NM (2009) Declining and low fecal corticoids are associated with distress, not acclimation to stress, during the translocation of African rhinoceros. Animal Conserv 13:104–111

Lukas KE (1999) A review of nutritional and motivational factors contributing to the performance of regurgitation and reingestion in captive lowland gorillas. Appl Animal Behav Sci 63:237–249

Lukas KE, Marr MJ, Maple TL (1998) Teaching operant conditioning at the zoo. Teach Psychol 25 (2):112–116

Lukas KE, Hoff MP, Maple TL (2003) Gorilla behavior in response to systematic alternation between zoo enclosures. J Appl Anim Behav Sci 81(4):367–386

Mackinnon J (1974) The behaviour and ecology of wild orangutans. Anim Behav 22:3–74

Mallapur A, Chellam R (2002) Environmental influences on stereotypy and the activity budget of Indian leopards in four zoos in southern India. Zoo Biol 21(6):585–595

Mann S (2005) Ethological farm programs and the "market" for animal welfare. J Agr Environ Ethics 18:369–382

Manrod JD, Hartdegen R, Burghardt GM (2008) Anim Cogn 11:267–273

Maple TL (1979a) Great apes in captivity: the good, the bad, and the ugly. In: Erwin J, Mitchell G, Maple T (eds) Captivity and behavior. Van Nostrand Reinhold, New York, pp 239–272
Maple TL (1979b) Primate psychology in historical perspective. In: Erwin et al. (eds) Captivity and behavior. Van Nostrand Reinhold, New York, pp 29–58
Maple TL (1980a) Orangutan behavior. Van Nostrand Reinhold, New York
Maple TL (1980b) Breaking the hand-rearing syndrome in captive apes. AAZPA regional proceedings. American Association of Zoological Parks and Aquariums, Wheeling WV, pp 199–201
Maple TL (1992) In memoriam: Prof. Dr. Heini Hediger (1908–1992). Zoo Biol 11:369–372
Maple TL (1996) The art and science of enrichment. In: Burghardt GM, Bielitzki JT, Boyce JR, Schaeffer DD (eds) The well-being of animals in zoo and aquarium sponsored research. Scientists Center for Animal Welfare, Greenbelt, MD, pp 79–84
Maple TL (2001) The power of one. In: Rabb G (ed) The apes: Challenges for the 21st century conferences proceedings. Chicago Zoological Society, Chicago, IL, pp 208–209
Maple TL (2005) Post-occupancy evaluation in the zoo: toward a science of appropriate, functional and superior exhibitory for animals and people. In: Plowman AB, Tonge SJ (eds) Innovation or replication? Proceedings of the 6th international symposium on zoo design. Whitley Wildlife Conservation Trust, Paignton Devon, UK, pp 111–118
Maple TL (2007) Toward a science of welfare for animals in the zoo. J Appl Anim Welf Sci 10(1):63–70
Maple TL (2008) Wildlife Wellness Center. Palm Beach Zoo Magazine III 2:10–11
Maple TL (2012) A zoo where the animals come first. Observer 39–41
Maple TL, Perkins LA (1995) Cage furnishings and environmental enrichment. In: Kleiman DG et al (eds) Wild mammals in captivity, vol 1. University Chicago Press, Chicago, IL, pp 212–222
Maple TL, Bashaw MJ (2010) Research trends in zoos. In: Kleiman DG et al. (eds) Wild mammals in captivity. University of Chicago Press, Chicago. Chapter 24, pp 288–298
Maple TL, Bloomstrand MA (1988) Feeding, foraging, and mental health. In: Meehan TP, Allen ME (eds) Proceedings of the third annual Dr. Scholl conference on the nutrition of wild animals. Lincoln Park Zoo Publication, Chicago, pp 1–8
Maple TL, Finlay TW (1986) Evaluating captive primate environments. In: Benirschke K (ed) Primates: The road to self-sustaining populations. Springer, New York, pp 480–488
Maple TL, Finlay TW (1987) Post-occupancy evaluation in the zoo. Appl Animal Behav Sci 18:5–18
Maple TL, Hoff MP (1982) Gorilla behavior. Van Nostrand Reinhold, New York
Maple TL, Kuhar CW (2006) The comparative psychology of Duane Rumbaugh and its impact on zoo biology. In: Washburn DA (ed) Primate perspectives on behavior and cognition. American Psychological Association, Washington, DC
Maple TL, Lindburg DG (eds) (2008) Empirical zoo: Opportunities and challenges to research in zoos and aquariums. Special Issue of Zoo Biology, 27(6) 431–504
Maple TL, Mallavarapu S (2006) Values, advocacy, and leadership: an empirical philosophy for 21st century zoos and aquariums. In: Genoways HH (ed) Museum philosophy for the 21st century. Rowman & Littlefield, Lanham, New York, Toronto, Oxford, pp 177–200
Maple TL, Stine WW (1982) Environmental variables and great ape husbandry. Am J Primatol 1 (Suppl):67–76
Maple TL, Erwin J, Mitchell G (1974) Sexually aroused self-aggression in a socialized adult male monkey. Arch Sex Behav 3(5):471–475
Maple TL, Zucker EL, Dennon MB (1979) Cyclic proceptivity in an adult female orangutan. Behav Processes 4:53–59
Maple TL, Bloomsmith MA, Martin AL (2009) Primates and pachyderms: a primate model of zoo elephant welfare. In: Forthman DL, Kane LF, Hancocks D, Waldau PF (eds) An elephant in the room: The science and well-being of elephants in captivity. Tufts University Center for Animals and Public Policy, North Grafton, MA

Margulis SW, Hoyos C, Anderson M (2003) Effect of felid activity on zoo visitor interest. Zoo Biol 22:587–599
Margulis SW, Rafacz M, Jacobs B (2005) Quantifying the effectiveness of environmental enrichment: lessons learned and rules of thumb. In: Clum N, Silver S, Thomas P (eds) Proceedings of the 7th ingternational conference on environmental enrichment. Wildlife Conservation Society, New York, pp 219–226
Markowitz H (1978) Engineering environments for behavioral opportunities in the zoo. Behavior Analyst, spring:34–47
Markowitz H (1979) Environmental enrichment and behavioral engineering for captive primates. In: Erwin J et al. (eds) Captivity and behavior. Van Nostrand Reinhold, New York, pp 217–228
Markowitz H (1982) Behavioral enrichment in the zoo. Van Nostrand Reinhold Co, New York
Markowitz H (2011) Enriching animal lives. Mauka Press, Pacifica, CA
Markowitz H, Aday C (1998) Power for captive animals: contingencies and nature. In: Shepherdson D et al (eds) Second nature: Environmental enrichment for captive animals. Smithsonian Institution Press, Washington and London, pp 47–58
Markowitz H, Aday C, Gavassi A (1995) Effectiveness of acoustic prey: environmental enrichment for a captive African leopard. Zoo Biol 14:371–379
Marr MJ (2006) Trends in international dissemination: behavior analysis on the world stage. Assoc Behav Anal Int Newsletter 29(3):16–17
Martin P, Bateson P (2007) Measuring behavior: An introductory guide, 3rd edn. University Press, Cambridge, UK
Martin AL, Bloomsmith MA, Kelley ME, Marr MJ, Maple TL (2011) Functional analysis and treatment of human-directed undesirable behavior exhibited by a captive chimpanzee. J Appl Behav Anal 44:139–143
Masefield W (1999) Forage preferences and enrichment in a group of captive Livingstone's fruit bats Pteropus livingstonii. Dodo 35:48–56
Masi S, Gustafsson E, Saint Jalme M, Narat V, Todd A, Bomsel M, Krief S (2012) Unusual feeding behavior in wild great apes, a window to understand origins of self-medication in humans: role of sociality and physiology in learning process. Physiol Behav 105(2):337–349
Maslow AH (1935) The role of dominance in the social and sexual behavior of infra-human primates: III. A theory of sexual behavior of infra-human primates. J Genet Psychol 48:310–338
Maslow AH (1936) The role of dominance in the social and sexual behavior of infra-human primates IV. J Genet Psychol 49:161–198
Maslow AH (1940) Dominance-quality and social behavior in infra-human primates. J Soc Psychol 11:313–324
Maslow AH (1962) Toward a psychology of being. Van Nostrand, New York
McMillan FD (2000) Quality of life in animals. J Am Vet Med Assoc 216:1904–1910
Mason WA (1965) The social development of monkeys and apes. In: DeVore I (ed) Primate behaviour. Holt, Rinehart and Winston, New York
Mason GJ (1991) Sterotypies: a critical review. Anim Behav 41:1015–37
Mason GJ, Burn CC (2011) Behavioural Restriction. In: Appelby MC, Mench JA, Olsson IAS, Hughes BO (eds) Animal welfare. CABI International, UK, pp 98–119
Mason GJ, Veasey JS (2010) How should the psychological well-being of zoo elephants be objectively investigated? Zoo Biol 29:237–255
Mason GJ, Clubb R, Latham N, Vickery S (2007) Why and how should we use environmental enrichment to tackle stereotypic behavior? Appl Animal Behav Sci 102:163–188
Matheson K (2012) Philadelphia Zoo: Trails Now Open for Animals to Explore. Huffington Post, August 23
McCowan B, Anderson K, Heagarty A, Cameron A (2008) Utility of social network analysis for primate behavioral management and well-being. J Appl Animal Behav Sci 109:396–405
McDonald S (1994) The Detroit Zoo chimpanzees: exhibit design, group composition, and the process of group formation. Int Zoo Yearbook 33:235–247
McElligott AG, Maggini I, Hunziker L, Konig B (2004) Interactions between red-billed oxpeckers and black rhinos in captivity. Zoo Biol 23:347–354

McGowan RTS, Robbins CT, Alldredge JR, Newberry RC (2010) Contrafreeloading in grizzly bears: implications for captive foraging enrichment. Zoo Biol 29:484–502
Melfi VA (2009) There are big gaps in our knowledge, and thus approach to zoo animal welfare: a case for evidence-based zoo animal management. Zoo Biol 28:574–588
Melfi VA, Hosey G (2011) Capacity building for better animal welfare. Int Zoo Yearbook 45:274–281
Melfi V, Bowkett A, Plowman A, Pullen K (2005) Do zoo designers know enough about animals? In: Plowman AB, Tonge SJ (eds) Innovation or replication? Sixth international symposium on zoo design. Paignton/Devon/UK, pp 119–227
Mellen J, Sevenich MacPhee M (2001) Philosophy of environmental enrichment: past, present, and future. Zoo Biol 20:211–226
Meller CL, Croney CC, Shepherdson D (2007) Effects of rubberized flooring on Asian elephant behavior in captivity. Zoo Biol 26:51–61
Meyer-Holzapfel M (1968) Abnormal behavior in zoo animals. In: Fox MW (ed) Abnormal behavior of animals, pp 476–503
Miller LJ (2012) Visitor reaction to pacing behavior: influence on the perception of animal care and interest in supporting zoological institutions. Zoo Biol 31:242–248
Mitchell G (1970) Abnormal behavior in primates. In: Rosenblum LA (ed) Primate behavior: Developments in field and laboratory research. Academic, New York, pp 195–249
Mitchell G, Obradovich S, Herring F, Dowd B, Tromborg C (1991) Threats to observers, keepers, visitors, and others by zoo mangabeys. Primates 32:515–522
Moberg GP (1985) Biological response to stress: key to assessment of animal well-being? In: Moberg GP (ed) Animal stress. Waverly, Baltimore, pp 27–49
Moore BD, Marsh KJ, Wallis IR, Foley WJ (2005) Taught by animals: how understanding diet selection leads to better zoo diets. Int Zoo Yearbook 39:43–61
Morris D (1964) The response of animals to a restricted environment. Sym Zoologic Soc London 13:99–118
Naald BV, Cameron TA (2011) Willingness to pay for other species well-being. Ecol Econ 70(7):1325–1335
Nasser HE (2012) Architects use design to help fight obesity. USA Today, October 28, 2012
Newberry RC (1995) Environmental enrichment: increasing the biological relevance of captive environments. Appl Animal Behav Sci 34:229–243
Nieuwenhuisen C, de Waal FBM (1982) Effects of spatial crowding on social behavior in a chimpanzee colony. Zoo Biol 1:6–28
Nishihara T (1992) A preliminary report on the feeding habits of western lowland gorillas in the Ndoki Forest, Northern Congo. In: Ityoigawa N et al (eds) Topics in primatology, behavior, ecology, and conservation, vol 2. University of Tokyo Press, Tokyo, Japan, pp 225–243
Nissen HW (1931) A field study of the chimpanzee: observations of chimpanzee behavior and environment in Western French Guinea. Comp Psychol Monographs 8:1–105
Norton BG, Hutchins M, Stevens E, Maple TL (eds) (1995) Ethics on the ark: Zoos, animal welfare and wildlife conservation. Smithsonian, Washington, DC, p 330
Novak M, Suomi SJ (1988) Psychological well-being of primates in captivity. Am Psychol 43:765–773
Nuzzo R (2005) Profile of frans B.M. de Waal. PNAS 102(32):11137–11139
Ogden JJ, Finlay TW, Maple TL (1990) Gorilla adaptations to naturalistic environments. Zoo Biol 9(2):107–121
Ogden JJ, Lindburg DG, Maple TL (1993) Preferences for structural environmental features in captive lowland gorillas. Zoo Biol 12(4):381–396
Ogden JJ, Lindburg DG, Maple TL (1994) A preliminary study on the effects of ecologically relevant sounds on the behavior of captive lowland gorillas. Appl Animal Behav Sci 39:163–176
Ogura T (2012) Use of video system and its effects on abnormal behavior in captive Japanese macaques (Macaca fuscata) Appl Animal Behav Sci
Oishi S, Diener E, Lucas RE (2007) The optimum level of well-being: Can people be too happy? Perspect Psychol Sci 2(4):346–360

Olney PJS (ed) (2005) Building a future for wildlife: The world zoo and aquarium conservation strategy. World Association for Zoos and Aquariums, Berne, Switzerland

Owen MA, Swaisgood RR, Czekala NM, Steinman K, Lindburg DG (2004) Monitoring stress in captive giant pandas: behavioral and hormonal responses to ambient noise. Zoo Biol 23:147–164

Paquet PCA, Darimont CT (2010) Wildlife conservation and animal welfare: two sides of the same coin. Animal Welfare 19:177–190

Parameswaren S (2012) Obese elephants in Tamil Nadu given slimming help. BBC News, Science and Environment Online, 14 September

Parr L (2001) Cognitive and physiological markers of emotional awareness in chimpanzees. Anim Cogn 4:223–229

Patrick P, Mathews C, Ayers D, Tunnicliffe S (2007) Conservation and education: prominent themes in zoo mission statements. J Environ Education 38(3):53–60

Payne KB, Langbauer WR, Thomas EM (1986) Infrasonic calls of the Asian elephant. Behav Ecol Socobiol 18:297–301

Perdue BM, Gaalema DE, Martin AL, Dampier SM, Maple TL (2010) Factors affecting aggression in a captive flock of Chilean flamingos. Zoo Biol 29:1–6

Perdue BM, Clay AW, Stoinski TS, Gaalema DE, Maple TL (2012) Technology at the zoo: The influence of a touchscreen computer on orangutans and zoo visitors. Zoo Biol 31:27–39

Perkins LA (1992) Variables that influence the activity of orangutans. Zoo Biol 11:177–186

Plowman AB, Tonge SJ (eds) (2005) Innovation or replication? Proceedings of the sixth international symposium on zoo design. Whitley Wildlife Conservation Trust, Paignton/Devon/UK

Pomerantz O, Terkel J (2009) Effects of positive reinforcement training techniques on the psychological welfare of zoo-housed chimpanzees (Pan troglodytes). Am J Primatol 71:687–695

Poole J, Granli P (2009) Mind and movement: Meeting the interests of Elephants. In: Forthman et al. (eds) An elephant in the room. Tufts Center for Animals and Public Policy, North Grafton, MA, pp 2–21

Popovich, DG, Dierenfeld ES (1997) Gorilla Nutrition. In: Ogden J, Wharton D (eds) Management of gorillas in captivity: Husbandry manual. Gorilla Species Survival Plan, Association of Zoos and Aquariums, Bethesda, MD

Pontzer H, Raichlin DA, Shumaker RW, ocobock C, Wich SA (2010) Metabolic adaptation for low energy throughput in orangutans. Proc Natl Acad Sci U S A 107:14048–14052

Powell DM, Carlstead K, Tarou LR, Brown JL, Monfort SL (2006) Effects of construction noise on behavior and cortisol levels in a pair of captive giant pandas. Zoo Biol 25(5):391–408

Priest GP (1991) Protecting primates: the psychology of animal care, part 2. Zoonooz 64:7–11

Pryor K (1981) The rhino likes violets. Psychology Today 92–98 April

Pryor K (1999) Don't shoot the dog. Bantam Books, New York

Pryor K (2009) Reaching the animal mind. Scribner, New York

Pryor K, Haag R, O'Reilly J (1969) The creative porpoise: training for novel behavior. J Exp Anal Behav 12:653–661

Rabb GB, Saunders CD (2005) The future of zoos and aquariums: conservation and caring. Int Zoo Yearbook 39:1–26

Read BW, Meier JE (1996) Neonatal care protocols. In: Kleiman et al. (eds) Wild mammals in captivity. University of Chicago Press, Chicago and London, pp 41–55

Reade LS, Waran NK (1996) The modern zoo: how do people perceive zoo animals? J Appl Animal Behav Sci 47:109–118

Rees PA (2009a) The sizes of elephant groups in zoos: implications for elephant welfare. J Appl Anim Welf Sci 12(1):44–60

Rees PA (2009b) Activity budgets and the relationship between feeding and stereotypic behaviors in Asian elephants in a zoo. Zoo Biol 28(2):79–97

Reinhardt V, Reinhardt A (1998) Environmental enrichment for caged rhesus macaques. Animal Welfare Institute, Washington, DC

Resende LS, Remy GL, Jr de Almeida Ramos V, Andriolo A (2009) The influence of feeding enrichment on the behavior of small felids (Carnivora: Felidae) in captivity. Zoo Biol 26(4):601–605
Riess BF, Ross S, Lyerly SB, Birch HG (1949) The behavior of two captive specimens of the lowland gorilla, Gorilla gorilla gorilla (Savage & Wyman). Zoologica 34(13):11–118
Rivas J, Burghardt GM (2002) Crotalomorphism: a metaphor for understanding anthropomorphism by omission. In: Bekoff M et al (eds) The cognitive animal: experimental and theoretical perspectives on animal cognition. MIT Press, Cambridge, MA, pp 9–17
Roediger R (2006) An outside perspective of contemporary behavior analysis: what happened to behaviorism? ABA Newsletter 29(3):21–24
Rogers CM, Davenport RK (1969) Effects of restricted rearing on sexual behavior of chimpanzees. Devel Psychol 1(3):200–204
Rose NA (2004) Captive cetaceans: the science behind the ethics. Paper presented at the european cetacean society conference. Kolmorden, Sweden
Ross SR, Bloomsmith MA (2011) A comment on Birkett and Newton-Fisher. PLoS One 6(6): e20101. doi:10.1371/journal.pone.0020101
Ross SR, Lukas KE (2006) Use of space in a non-naturalistic environment by chimpanzees and lowland gorillas. Appl Animal Behav Sci 96(1–2):143–152
Ross SR, Lukas KE, Lonsdorf EV, Stoinski TS, Hare B, Shumaker R, Goodall J (2008) Inappropriate use and portrayal of chimpanzees. Science 319:1487
Ross SR, Shapiro SJ, Hau J, Lukas KE (2009) Space use as an indicator of enclosure appropriateness: A novel measure of captive animal welfare. Appl Animal Behav Sci 121:42–50
Rumbaugh DM, Washburn DA (2003) Intelligence of apes and other rational beings. Yale University Press, New Haven
Sargent TD, Keiper R (1967) Stereotypies in caged canaries. Anim Behav 15(1):62–66
Savastano G, Hanson A, McCann C (2003) The development of an operant conditioning program for new world primates at the Bronx Zoo. J Appl Anim Welf Sci 6(3):247–261
Schaul J (2012) A critical look at the future of zoos—an interview with David Hancocks. Newswatch/National Geographic.com, March
Segal EF (ed) (1989) Housing, care and psychological wellbeing of captive and laboratory primates. Noyes Publications, Park Ridge, NJ
Seidensticker J, Doherty JG (1996) Integrating behavior and exhibit design. In: Kleiman DG, Allen ME, Thompson KV, Lumpkin S (eds) Wild mammals in captivity. University of Chicago Press, Chicago and London, pp 191–203
Seiffert S (2009) Developing and sharing information about the use of plants within animal collections: the BIAZA zoo plant wiki. ICEE Conference Proceedings, Paignton Zoo, Torquay, UK
Seligman MEP, Czikszentmihalyi M (2000) Positive psychology: an introduction. Am Psychol 55 (1):5–14
Shepherdson DJ (1989) Improving animals lives in captivity through environmental enrichment. In: Close BS, Dolins F, Mason GJ (eds) Euroniche conference proceedings, Edinburgh, Scotland. Humane Education Center, London, UK, pp 91–102
Shepherdson DJ, Carlstead K, Mellen JD, Seidensticker J (1993) The influence of food presentation on the behavior of small cats in confined environments. Zoo Biol 12:203–216
Shepherdson D, Mellen JD, Hutchins M (eds) (1998) Second nature: Environmental enrichment for captive animals. Smithsonian Institution Press, Washington and London
Shumaker RW, Walkup KR, Beck BB (2011) Animal tool behavior. Johns Hopkins University Press, Baltimore, MD
Shyne A (2006) Meta-analytic review of the effects of enrichment on stereotyped behavior in zoo mammals. Zoo Biol 25(4):317–338
Siebert C (2006) The Animal Self. New York Times.com. pp 12
Silk JB, Beehner JC, Bergman TJ, Crockford C, Engh AL, Moscovice LR, Wittig RM, Seyfarth RM, Cheney DL (2010) Strong and consistent social bonds extend the longevity of female baboons. Curr Biol 20(15):1359–1361

Snyder RJ, Perdue BM, Powell D, Forthman DL, Bloomsmith MA, Maple TL (2012) Behavioral and hormonal consequences of transporting giant pandas from China to the United States. J Appl Anim Welf Sci 15:1–20

Soltis J, Brown JL (2010) Special issue on the care and welfare of elephants in AZA institutions. Zoo Biol 29(2):85–283

Soltis J, Leighty KA, Wesolek CM, Savage A (2009) The expression of affect in African elephant (Loxodonta africana) rumble volcaizations. J Comp Psychol 123:222–225

Sommer R (1972) What do we learn at the zoo? Natural History 81(7):26–29–84–85

Sommer R (1974) Tight spaces. Prentice-Hall, Englewood Cliffs, NJ

Sommer R (2008) Development of the zoological garden and mental hospital. Am J Orthopsychiatry 78(3):378–382

Sommerfeld R, Bauert M, Hillmann E, Stauffacher M (2006) Feeding enrichment by self-operated food boxes for white-fronted lemurs in the Masoala exhibit of the Zurich Zoo. Zoo Biol 25:145–154

Steele J (2009) Room to roam: San Diego Zoo hopes bigger elephant space will improve the animals' health, numbers. San Diego Union-Tribune, January 25, 2 pp

Stevens EF, Pickett C (1994) Managing the social environments for flamingos for reproductive success. Zoo Biol 13:501–507

Stoinski TS, Daniel E, Maple TL (2000) A preliminary study of the behavioral effects of feeding enrichment on African elephants. Zoo Biol 19:485–493

Stoinski TS, Hoff MP, Lukas KE, Maple TL (2001) A behavioral comparison of two captive all-male gorilla groups. Zoo Biol 20(1):27–40

Stoinski TS, Allen MT, Bloomsmith MA, Forthman DL, Maple TL (2002) Educating zoo visitors about complex environmental issues: should we do it and how? Curator 45(2):129–143

Stoinski TS, Lukas KE, Kuhar CW, Maple TL (2004) Factors influencing the formation and maintenance of all-male gorilla groups in captivity. Zoo Biol 23(2):189–203

Stoinski TS, Jaicks HF, Drayton LA (2011) Visitor effects on the behavior of captive western lowland gorillas: The importance of individual differences in examining welfare. Zoo Biol 30:1–14

Swaisgood RR (2007) Current status and future directions of applied behavioral research for animal welfare and conservation. Appl Animal Behav Sci 102:139–162

Swaisgood RR, Ellis S, Forthman DL, Shepherdson J (2003) Improving well-being for captive giant pandas: theoretical and practical issues. Zoo Biol 22:347–354

Swaisgood RR, Shepherdson DJ (2005) Scientific approaches to enrichment and stereotypies in zoo animals: what's been done and where should we go next? Zoo Biol 24(6):499–518

Swaisgood RR, Ellis S, Forthman DLO, Shepherdson DJ (2003) Improving well-being for captive giant pandas: theoretical and practical issues. Zoo Biol 22:347–354

Szokalski MS, Litchfield CA, Foster WK (2012) What can zookeepers tell us about interacting with big cats in captivity? Zoo Biol 31:1–10

Tudge C (2001) Animal experiences. Lond Rev Books 23(12):38–39

Tarou LR, Bashaw MJ (2007) Maximizing the effectiveness of environmental enrichment: Suggestions from the experimental analysis of behavior. Appl Animal Behav Sci 102:189–204

Tarou LR, Snyder RJ, Maple TL (2004) Spatial memory in the giant panda. In: Lindburg DG, Baragona K (eds) Giant pandas: Biology and conservation. University of California Press, Berkeley, pp 125–132

Tavolga MC (1966) Behavior of bottle-nose dolphins (Tursiops truncatus): social interaction in a captive colony. In: Norris KS (ed) Whales, dolphins, and porpoises. University of California Press, Berkeley, pp 718–730

Thelen E (1981) Rhythmical behavior in infancy: an ethological perspective. Dev Psychol 17(3):237–257

Thompson V (1976) Observation of great apes in a naturalistic zoo environment. Lincoln Park Zoo Publication, Chicago

Tovar TC, Moore D, Dierenfield E (2005) Preferences among four species of local browse offered to Colobus guereza kikuyuensis at the Central Park Zoo. Zoo Biol 24:267–274
Tresz H (2006) Behavioral management at the Phoenix Zoo: New strategies and perspectives. J Appl Anim Welf Sci 9(1):65–70
Tutin CEG, Fernandez M (1993) Composition of the diet of chimpanzees and comparisons with that of sympatric lowland gorillas in the Lope Reserve, Gabon. Am J Primatol 30:195–211
Van Hoof JARAM (1973) The Arnhem Zoo chimpanzee consortium: an attempt to create an ecologically and socially acceptable habitat. Int Zoo Yearbook 13:195–203
Vander Naald B, Cameron TA (2011) Willingness to pay for other species' well-being. Ecol Econ 70:1325–1335
Veasey J (2005) Whose zoo is it anyway? Integrating animal, human and institutional requirements in exhibit design. In: Plowman AG, Tonge SJ (eds) Innovation or replication? Proceedings of the sixth international symposium on zoo design, Paignton/Devon/UK, pp 7–16
Washburn SL, DeVore I (1961) The social life of baboons. Scientific American 204:62–72
Washburn DA, Rumbaugh DM (1992) Investigations of rhesus monkey video-task performance: Evidence for enrichment. Contemp Top Lab Anim Sci 31(5):6–10
Wasser SK, Hunt KE, Brown JL, Cooper K, Crockett CM, Bechert U, Millspaugh JJ, Larson S, Monfort SL (2000) Generalized fecal glucocorticoid assay for use in a diverse array of nondomestic mammalian and avian species. Gen Comp Endocrinol 120:260–275
Watters J (2009) Toward a predictive theory for environmental enrichment. Zoo Biol 28(6):609–622
Watters J, Wielebnowski N (2009) Special issue on zoo animal welfare. Zoo Biol 28(6):501–651
Weir K (2011) Evidence is mounting for the benefits of exercise, yet psychologists don't often use exercise as part of their treatment arsenal. Here's more research on why they should. Monit Psychol, 42(11):49–52
Wells DL, Irwin RM (2008) Auditory stimulation as enrichment for zoo-housed Asian elephants (Elephas maximus). Animal Welfare 17:335–340
Wells DL, Coleman D, Challis MG (2005) A note on the effect of auditory stimulation on the behavior and welfare of zoo-housed gorillas. Appl Animal Behav Sci 100:327–332
Wemmer C, Christen CA (2008) (eds) Elephants and ethics: Toward a morality of coexistence. Johns Hopkins University Press, Baltimore
White M (1981) What is and what ought to be done: An essay on ethics and epistemology. Oxford University Press, New York
White BC, Houser LA, Fuller JA, Taylor S, Elliott JLL (2003) Activity-based exhibition of five mammalian species: evaluation of behavioral changes. Zoo Biol 22:269–285
Whitham JC, Wielebnowski N (2009) Animal-based welfare monitoring: using keeper ratings as an assessment tool. Zoo Biol 28:545–560
Whittaker M, Laule G (2009) Protected contact and elephant welfare. In: Forthman DL et al (eds) An elephant in the room. Center for Animals and Public Policy, Tufts University, North Grafton, MA, pp 181–188
Wielebnowski NC, Fletchall N, Carlstead K, Busso JM, Brown JL (2002a) Noninvasive assessment of adrenal activity associated with husbandry and behavioral factors in the North American clouded leopard population. Zoo Biol 21:77–98
Wielebnowski NC, Ziegler K, Wildt DE, Lukas J, Brown JL (2002b) Impact of social management on reproductive, adrenal and behavioural activity in the cheetah (Acinonyx jubatus). Animal Conserv 5:291–301
Wiese RJ, Willis K (2004) Calculation of longevity and life expectancy in captive elephants. Zoo Biol 23:365–373
Wilson SF (1982) Environmental influences on the activity of captive apes. Zoo Biol 1:201–209
Wilson M, Kelling A, Poline I, Bloomsmith MA, Maple TL (2003) Post-occupancy evaluation of Zoo Atlanta's giant panda conservation center: staff and visitor reactions. Zoo Biol 22 (4):365–382

Wilson M, Bashaw MF, Fountain K, Kieschnick S, Maple TL (2006) Nocturnal behavior in a group of female African elephants. Zoo Biol 25(3):173–186

Wineman J, Choi YK (1991) Spatial/visual properties of zoo exhibition. Curator 34(4):304–315

Wortman JD, LarRue MD (1974) Hand-rearing polar bear cubs Thalarctos maritimus at Topeka Zoo. Int Zoo Yearbook 14:215–218

Wuichet J, Norton BG (1995) Differing conceptions of animal welfare. In: Norton BG et al (eds) Ethics on the ark. Smithsonian, London and Washington, pp 235–250

Yamanishi Y, Hayashi M (2011) Assessing cognitive experiments on the welfare of captive chimpanzees by direct comparison of activity budget between wild and captive chimpanzees. Am J Primatol 73(12):1231–1238

Yeates JW, Main DCJ (2008) Assessment of positive welfare: a review. Vet J 175:293–300

Yerkes RM (1916) The mental life of monkeys and apes: a study of ideational behavior. Behav Monographs 3:1–145

Yerkes RM (1925) Almost human. The Century Company, London/New York

Yerkes RM (1927) The mind of a gorilla. Genet Psychol Monogr 2:1–193

Yerkes RM (1943) Chimpanzees: A laboratory colony. Yale University Press, New Haven

Yerkes RM, Yerkes AW (1929) The great apes. Yale University Press, New Haven

Young RJ (2003) Environmental enrichment for captive animals. Blackwell, Oxford

Index

DOI 10.1007/978-3-642-35955-2,

The manufacturer's authorised representative in the EU is Springer
Nature Customer Service Centre GmbH, Europaplatz 3, 69115 Heidelberg,
Germany. If you have any concerns regarding our products, please
contact ProductSafety@springernature.com

Printed and bound by CPI Group (UK) Ltd, Croydon, CR0 4YY
30/09/2026
02225591-0007